❖ VERIFYING NONPROLIFERATION TREATIES

Obligation, Process, and Sovereignty

❖ VERIFYING NONPROLIFERATION TREATIES

Obligation, Process, and Sovereignty

J. Christian Kessler

National Defense University Press
Washington, DC

National Defense University Press Publications

To increase general knowledge and inform discussion, the Institute for National Strategic Studies, through its publication arm the NDU Press, publishes McNair Papers; proceedings of University- and Institute-sponsored symposia; books relating to U.S. national security, especially to issues of joint, combined, or coalition warfare, peacekeeping operations, and national strategy; and a variety of briefer works designed to circulate contemporary comment and offer alternatives to current policy. The Press occasionally publishes out-of-print defense classics, historical works, and other especially timely or distinguished writing on national security.

NDU Press publications are sold by the U.S. Government Printing Office. For ordering information, call (202) 512-1800 or write to the Superintendent of Documents, U.S. Government Printing Office, Washington, DC 20402.

Library of Congress Cataloging-in-Publication Data

Kessler, J. Christian

Verifying nonproliferation treaties: obligation, process, and sovereignty / J. Christian Kessler.

p. cm.

Includes bibliographical references.

1. Nuclear arms control—Verification. 2. Chemical arms control—Verification. 3. Biological arms controls—Verification. I. Title

JX1974.K45

327.1'74—dc20 95-41341

CIP

First Printing, October 1995

For sale by the U.S. Government Printing Office
Superintendent of Documents, Mail Stop: SSOP, Washington, DC 20402-9328
ISBN 0-16-051683-8

Contents

❖ Foreword

This book examines two of the most critical issues of national security policy: stemming the proliferation of weapons of mass destruction, and verifying international nonproliferation agreements. Contrary to predictions made in the 1970s, there are relatively few countries today that either possess or are pursuing nuclear weapons, although there are some pursuing chemical and biological weapons. This general success is a result of several most unusual international treaties—in the cases of chemical and biological weapons, treaties banning such weapons outright, and in the case of nuclear weapons, a treaty limiting the authorized possession of such weapons.

If the integrity and effectiveness of these treaties are to be sustained, the question to be addressed is how each party maintains confidence that other parties are abiding by the obligations each has undertaken. It is the character of this verification that has been the subject of considerable debate, with certain critics arguing that real verification is unachievable. Whatever the prospect for successful verification, there is widespread agreement that modern treaties for the purpose of enhancing national security must be verifiable in some fashion; no government should take the good behavior of its adversary as an article of faith. As President Reagan said, "Trust but verify."

Mr. Kessler's book speaks to the student as well as the practitioner. To the student he describes the forces that shape the ways in which negotiators address the question of establishing and maintaining confidence among signatories to treaties limiting nuclear, chemical, and biological weapons and reducing the size of conventional forces. To the practitioner he explains how the chance events of the day—fate and circumstance—intertwine with substantive developments in treaty negotiations and implementation. To both audiences he speaks from 15 years of active experience in studying and negotiating how to limit the dangers of weapons of mass destruction.

ERVIN J. ROKKE
Lieutenant General, USAF
President, National Defense University

❖ Acknowledgments

I wish to note the support I received from a number of individuals whose guidance and criticism played an important role in the final product. Ambassador Robert Joseph acted as mentor for this enterprise, providing particular insight and guidance on the chemical and biological regimes as well as access to numerous government officials whose insights were critical to the success of this project. Dr. Thomas Keaney acted as friend and confessor and read each draft of each chapter with the eye of the nonexpert, asking repeated questions about the emperor's wardrobe, or lack thereof. Detailed comments on the manuscript from Dr. Leonard Spector, Mr. Charles Van Doren, Mr. David Kay, and several diligent and selfless officers in the Arms Control and Disarmament Agency's International Verification Bureau made significant contributions to the final shape of my arguments.

❖ VERIFYING NONPROLIFERATION TREATIES

Obligation, Process, and Sovereignty

❖ The Framework

Throughout history sovereigns have faced the same problem: How can one sovereign be confident that another will abide by or fulfill an agreement? Over the long span of history, most sovereigns have been individuals—princes, kings, or emperors—and they have addressed this question in a manner appropriate for individuals, such as by exchanging "hostages," family members or other subjects personally important to the sovereign, or by exchanging some other form of earnest.

During the 17th century, the concept of sovereigns as individual monarches gave way to the Westphalian state system of governments as sovereigns. In this system, states are independent actors answering to no higher authority, where earlier even the monarch had answered to God and, until the Reformation, the Pope. Under this new system, because each state stood alone and independent, it was and could be the only protector of its own interests. During the 17th and 18th centuries the notion developed that a state would not freely enter into an agreement not beneficial to it; mutual advantage was the basis for having confidence in the intention of the other party (or parties) to comply with the commitments made.

This notion that a sovereign would enter into an agreement only if it were beneficial to it, abiding by the agreement because and only as long as it was beneficial, remained prevalent until after World War I. There were, of course, recognized limits for compliance implied by this doctrine, and these were essentially those identified some 400 years earlier as appropriate for princes and other individual monarchs: "A prudent ruler ought not to keep faith when by so doing it would be against his interests, and when the reasons which made him bind himself no longer exist."[1] Such a doctrine implies the need to include in each agreement some mechanism to maintain the interest of other sovereigns in complying with it. This doctrine can also be taken to imply that the sovereign is supreme; hence accepting constraints on his behavior would be inappropriate. For the individual sovereign, it seems that accepting checks on his good faith was simply another element of the exercise of sovereignty. For the corporate sovereigns of the Westphalian system, accepting such constraints came to be seen as implying a diminution of power rather than its exercise.[2] Nonetheless, governments gradually came to recognize, in the course of negotiating particular agreements, that each of the parties would benefit if all agreed to some mechanism for judging the compliance of the other parties.

From such efforts the modern concept of "verification" has developed—a procedure for evaluating the compliance of each party with the agreement made. Verification may be performed by each of the parties with respect to the others (as it was traditionally, to the extent that verification in its modern sense was performed at all). Alternatively, verification may be performed by some neutral entity, commonly an international organization comprising the states party to a convention. (While other arrangements can be imagined, they have not been attempted in practice except in relatively local or minor agreements.)

The evolution of verification has seen its greatest expression in treaties dealing with weapons capable of conveying the ultimate military advantage—nuclear, chemical, and biological weapons. These "weapons of mass destruction" create the potential for any state, even a militarily powerful one, to be destroyed by an adversary able to achieve strategic surprise in using them. For this reason, multilateral treaties controlling or eliminating these weapons have been on the international agenda for most of the 20th century, and verification measures—and ways to ensure their effectiveness— have been a central issue in the negotiation of these agreements.

Today, the effectiveness of such agreements in preventing the proliferation of nuclear, chemical, and biological weapons is a central concern in foreign affairs and national security policy. This study examines the factors that shaped the verification systems in the three universal[3] nonproliferation conventions:

- The Treaty on the Non-Proliferation of Nuclear Weapons (NPT)
- The Biological Weapons Convention (BWC),
- The Chemical Weapons Convention (CWC).

The evolution of each of these multilateral agreements is considered, and the Conventional Forces in Europe Treaty (CFE) is also examined, to provide a "control case"—a case similar in many of the structural factors shaping it but one with a very different verification system.

To comprehend the significance of particular choices among alternative approaches to verification, it is essential to understand better why states bother with this issue at all. It is necessary to look at the question of what is meant by the term "verification" in modern usage and consider in greater depth why states frequently want verification measures included in multilateral agreements; then examine the verification provisions in each of the four agreements and trace the history of how those particular provisions came about; and, finally, evaluate the degree to which the respective verification systems were shaped by specific substantive factors and the degree to which random historical events are the more significant causal factors.

Notes

1. Niccolo Machiavelli, *The Prince*, trans. Luigi Ricci, rev. ed. H.R.P. Vincent (New York: The New American Library, 1952), 92.

2. The reason for this development is well beyond the scope of the current study, but it is intriguing to note the degree to which one theme runs strongly and consistently through the negotiation of the IAEA Statute, the NPT, the BWC, and the CWC: accepting inspections by an international inspectorate to verify compliance is frequently seen in the modern world as a diminution of sovereignty rather than an exercise of that sovereignty. This position has been argued by countries as historically different as China, India, the Soviet Union, and Belgium.

3. The term "universal" is used here and in the following chapters to indicate treaties open to all states, as opposed to treaties which can be adhered to only by states meeting some specific criterion, normally location with a specified region. No treaty enjoys the status of universal adherence in the literal sense of all states in the international system, even the United Nations Charter.

1. ❖ Verification: The Process

Verification is a frequently misunderstood subject, one reason being that verification can be highly technical, and many of the critical details for any particular verification system depend on the specific technologies involved in the activities to be examined. Verifying a prohibition on chemical weapons, for example, involves many technical issues unique to the production of specific chemicals and to the requirements of converting those chemicals to weapons. Verifying a prohibition on nuclear weapons is characterized by a different set of technical issues; verifying the elimination (or the limitation) of a specific class of weapons systems (such as the Intermediate Nuclear Forces agreement) will involve still other technical matters.

Verifying a state's compliance with its treaty commitments has a certain inherent logic and structure common to all situations, regardless of the technical aspects. This underlying logic reflects a second reason for so much misunderstanding—verification touches on fundamental security issues for countries considering participation in a particular treaty regime, because it addresses whether other parties are abiding by their commitments or are cheating. Hence a great deal of political emotion frequently accompanies discussions of verification and its adequacy. Nonetheless, it has become a central element of international agreements in many fields beyond classical national security.[1]

What is Verification?

Long an important question in arms control and disarmament discussions, verification is considered by many today to be a critical element of any effort to control or eliminate weapons of mass destruction. In this view, verification is necessary to provide international security with respect to peaceful uses of technologies potentially useful for weapons of mass destruction, but there are others who perceive verification as the unattainable aspiration that can make an arms control or disarmament treaty more probably a Trojan horse than an effective instrument for security. So, what is verification and what role does it play in the overall question of implementing the objectives of an international agreement to control the spread of one kind of weapon of mass destruction?

To understand the nature of verification and its importance in treaties controlling or eliminating (at least in intent) weapons of mass destruction, it is important to understand the context in which the issue of verification arises. This context includes the factors that motivate governments to seek such weapons in the first place, and the architecture of international regimes that governments collectively create to resolve the problem.

There are three reasons why a country might choose to acquire weapons of mass destruction:[2]

- The country may face a significant threat to its security that cannot be met by conventional forces because of some strategic imbalance, which might arise because the threatening country is substantially larger, substantially more technologically advanced, or itself has weapons of mass destruction. Examples include China, Pakistan, and Israel.
- The country may have hegemonic or imperial designs that will be facilitated by (known) possession of weapons of mass destruction. Iraq, and probably Iran, have been motivated largely by hegemonic aspirations.
- The motivation for acquiring weapons of mass destruction may simply reflect a desire to establish bona fides as a major power. France is one example; it also seems to have been the basis for Argentine and Brazilian aspirations for nuclear weapons.[3]

For the state confronting a strategic imbalance, the situation can be ameliorated (but not entirely resolved) by effective (regional or global) collective security agreements, defensive alliances, and regional or bilateral confidence-building measures. However, a would-be hegemon cannot be controlled in the same fashion. The proliferant tendencies of countries fitting each situation can only be dealt with through an effective regime to control the spread of the particular technology plus effective regional or global collective security or defensive alliance arrangements.

It is for this reason that proliferation experts think in terms of building control regimes. A control regime is a fabric of international legal requirements reflecting and/or establishing accepted norms of national behavior, and mechanisms to implement or operationalize these requirements. In the ideal case, an international control regime includes a number of major elements, the first of which is an international treaty committing each state party to foreswear acquisition, possession, use, or threat of use of the subject weapon of mass destruction.[4] Normally the prohibitions are absolute, but in the case of nuclear weapons, the Treaty on the Non-Proliferation of Nuclear Weapons (NPT) permits the five states that had already acquired nuclear weapons (and proven this by conducting tests) to maintain their arsenals, while prohibiting any further states parties from acquiring or possessing nuclear weapons. The treaty commitment is tailored to the subject and the political situation at the time the agreement is concluded. As we shall see in the following chapters, the character of the treaty commitment plays a large role in the character of the regime established to verify that commitment.

The second element is an international agreement (or, more rarely, a semiformalized consensus) mandating national controls on trade in items relevant to developing the weapon to be controlled or prohibited. In the nuclear case, consensus is reflected in letters written from national authorities to the Director General of the IAEA, while in the Chemical Weapons Convention it is reflected in the treaty itself.[5]

The third ideal element is verification, a mechanism or procedure to confirm that each state party to the agreement is acting in conformity with its obligations, and to detect those who violate their obligations. In modern multilateral treaties, the legal mandate to perform this confirmation activity is frequently given to an international organization created specifically or even solely for that purpose. In other cases it is left (as was the practice historically) to each state to perform individually. Frequently, the only practical and reasonably effective means of collecting the information necessary for unilateral verification is intelligence activities. Even in cases where a multilateral verification system is established, some states will choose to conduct their own information collection and evaluation activities in addition. For example, it has long been U.S. practice to collect and evaluate information about whether certain states, whose intentions the United States holds in doubt, are in fact complying with their treaty obligations, even when a multilateral compliance verification system exists.[6]

A final element is sanctions, some means for imposing pressure to conform on regime violators. Some regimes include specific sanctions relating to cooperation within the regime, while other regimes do not. In all cases, the ultimate sanction is reference to the U.N. Security Council and resulting action under Chapter VII of the Charter (which, ironically, exists largely independent of the particular regime; threats to international peace and security need not also be violations of a multilateral treaty, although it strengthens the case).

Each of these ideal elements could be, and has been repeatedly been, the subject of a separate book. The focus of the following pages is the process of confirming that each state party to a nonproliferation or arms control agreement is complying with its obligations under that agreement—"verification." As the term is used here and generally in the community concerned with controlling the proliferation of different weapons of mass destruction, verification refers to a set of activities. Again, in the ideal case, verification would be expected to include:

- National declarations of items (materials, equipment, etc.) and facilities which could be of use in conducting the forbidden activity, but which are for related but legitimate activities
- Onsite inspection by international or foreign inspectors
- Monitoring of items, areas, or activities
- Evaluation of the information generated by the above activities
- Judgments that, on the basis of available information, the inspected party has or has not fulfilled its commitments during some time period.

The essential predicate for verification is a formal commitment (adherence to a multilateral treaty in most cases) not to engage in specified activities, frequently with the explicit understanding that some set of related activities is not only permissible but expected (a prohibition of nuclear, biological, or chemical weapons, but conduct of a civil nuclear, bio-technology, or chemical industry). Verification then consists of several analytically and practically distinct activities:

Collecting information on relevant activities of each state party (which should include but may not be limited exclusively to that set of permissible activities related to the prohibited ones); evaluating that information to consider whether and to what degree there is evidence which may indicate that the state is violating its commitments; and making a judgment (a decision based on the weight of the evidence) as to whether the state party is or is not complying with its commitments. Policy analysts frequently think about this activity in terms of confirming compliance, but approached analytically, the activity must boil down to seeking to determine whether there is credible evidence indicating that a state is engaged in activities violating its commitments. Verification is, in essence, seeking to prove the negative, to prove that certain behavior does not exist.

Proving the nonexistence of prohibited behavior is inherently problematic and must be inductive rather than deductive. Logically one can only demonstrate the failure to detect proscribed behavior, one cannot demonstrate that it does not exist. In addition to this logical uncertainty, there are of course additional sources of uncertainty introduced at every level of activity in verification. Some of these uncertainties may be introduced with the technique of random sampling, and hence be reasonably understood. Other uncertainties may reflect factors that cannot be known or even which have not been identified.[7]

These uncertainties apply equally to unilateral and multilateral efforts to judge treaty compliance. There are several reasons behind a multilateral system of verification. One is to provide all parties a substantial capability to detect covert noncompliance by spreading the cost and making it affordable. A second is to raise the ante for covert violation, in the sense that because the probability of detection is increased, the magnitude of effort (cost and complexity) of successful covert violation is likewise significantly increased. A former Inspector General of the IAEA likened international nuclear safeguards to the European development of the handshake—extending one's sword hand demonstrates peaceful intention by decreasing offensive potential and increasing vulnerability. But as medieval men of noble standing wore both a sword *and* a dirk, accessible to the left hand, the handshake is a significant gesture to indicate peaceful intentions, but it does not make one absolutely incapable of surprise attack. Both "disarming" and visible evidence of good intentions can provide an increased level of confidence, but not certainty.

It is for these reasons that, while verification is an important component of an international nonproliferatin regime, it is but one component. Even when verification may have a high probability of detecting violations, additional measures are required. Some, like export controls, are to make violation more difficult by limiting access to the materials and equipment needed to commit violations. Others, like continued national intelligence collection, can provide information on specific behaviors that indicate a change in intentions, or even just a change in how another state party perceives the threat environment from that which existed when the original nonproliferation undertaking was made. National intelligence may also detect acquisition by covert means of the materials, equipment, and technology relevant to developing the weapon of mass destruction in violation of treaty commitments. (This acquisition may involve diversion of items imported ostensibly for legitimate peaceful purposes, or may be indigenous, or may even involve acquisition of items lower in the process chain than controlled items, but which are acquired in a covert fashion or for which the state has no peaceful use.) Intelligence or investigation may also detect the development of or training in military doctrines or capabilities related to deployment and use of the weapon.

One important function of international nuclear safeguards is making legitimate activities concerning the peaceful use of nuclear energy transparent to the international community. To the extent that a facility (or program or activity) is well known and open to international visitors, concerns about its purpose are reduced. Like any confidence-building measure, international transparency in nuclear activities is convincing only to the extent that concerns about intentions are low or moderate. If a state believes that its neighbor is dedicated to obtaining weapons of mass destruction, there is a presumption that it will take all affordable and technically feasible steps to conceal its real activities and attempt to use confidence-building measures to obscure the truth.

Perhaps for this reason, and because technical sources of uncertainty are often not appreciated, commentators frequently draw a sharp distinction between confidence-building measures and verification. In this dichotomy, confidence-building measures involve actions that manifest peaceful intentions and good faith but that all agree do not constitute proof.[8] Confidence-building measures can reduce tensions and begin to build bridges between adversaries, but only to the extent that there is a predisposition to do so. Confidence-building measures are impotent in the face of deep distrust and will be viewed as efforts to mislead and conceal offensive efforts instead of as efforts to demonstrate good faith. In contrast, verification is, almost by definition, expected to provide some high degree of confidence—certainty is a word frequently used—that foresworn behaviors are not being conducted covertly. By this definition, verification can be relied upon in the face of great distrust, and the inspected state given a "clean bill of health," but if distrust is so great, what level of confidence that violations are detected will be adequate? Discussions of whether a particular arms control or nonproliferation agreement is in the national interest always seem to find some who argue that the achievable level of detection is inadequate.

As the foregoing discussion has begun to show, uncertainty is inherent in verification, and in some respects even the level of uncertainty (a formal concept in statistics, but not in conventional thought) is uncertain. The dichotomy between "confidence-building measures" and "verification" measures is in fact vague and of limited utility. All verification measures are in some sense confidence-building measures, although the latter make up a significantly larger conceptual category, and "verification" is specifically limited to measures providing evidence and a judgment that the state is *not* engaging in activities violating a treaty commitment. However, as described above, verification cannot prove the negative, which is a logical impossibility. For this reason alone, no practical set of verification measures can provide absolute assurance. Technological constraints are simply additional, secondary reasons why there can be no absolute certainty through verification. Notwithstanding these logical and technical constraints, one can collect information which significantly increases the probability that a judgment about the behavior of another state is accurate.

Why Verify?

But all of this begs the question, why verify? If certainty regarding the compliance of others were possible at a reasonable price, then the answer would be obvious—but we have just learned that certainty is not possible, and we have not yet addressed the question of cost. Why, in the face of such uncertainties and costs, do governments agree to limitations, frequently limitations imposing significant restrictions on the options available for protecting their national security? One interesting model, offered by Julian Perry Robinson[9], is utilitarian and involves the concept of two "baskets"—two broad policy alternatives, each with elements and aspects that are variable. Neither basket offers any absolutes, and each offers means for addressing problems of protecting the national security.

The first basket is unilateral action. In this case, the government undertakes no obligations with respect to other governments and accepts no restrictions on its actions save those imposed by its own national capabilities. The security of the state depends on the abilities of the government to determine both the intentions and capabilities of its neighbors,[10] and to obtain and use the diplomatic and military means to counter all threats. To the degree that some particular weapon appears useful for meeting national objectives, the government will evaluate whether it is affordable. In some cases, the weapon may have no value for meeting national objectives but must be considered as part of the offensive arsenal of adversaries. Thus, for many countries, chemical or biological weapons may provide no direct benefit, but the possibility of needing to protect against them remains. This is true whether the state confronts a rival and the potential for hostilities is high, or whether the threat environment is relatively benign. The degree of threat varies, but in either case the government must consider both the cost of effective countermeasures (either defensive systems or maintenance of a credible deterrent) and the costs of failing to possess such countermeasures when the need arises.

In an uncertain environment, any measure that can reduce the likelihood of threats deserves attention, and for this reason governments have since time begun sought agreements to reduce threats to their security. One modern example of immediate relevance is the Geneva Protocol for the Prohibition of the Use in War of Asphyxiating, Poisonous or Other Gases, and of Bacteriological Methods of Warfare. This agreement, in one page, proclaims that chemical weapons are contrary to customary international law and obligates parties to foreswear the first use in war of chemical weapons and biological ("bacteriological") weapons against other parties. Such an international agreement establishes a norm of behavior and gives all states some confidence that any state that breaks it will thereby gain the opprobrium of the international community. Opprobrium does little to protect the national security, however. While such international agreements can to some degree reduce the likelihood of chemical or biological attack, they will change the calculus only in the most marginal of cases. Any state with more than the most minimal of potential adversaries must still maintain countermeasures.

Let us assume an international norm prohibiting some behavior, established or codified in a treaty considerably more effective than the Geneva Protocol proved to be. Absent some international mechanism to verify compliance, the only means for detecting violations is for one state to collect information and to accuse another. This means that detecting violations is left to adversaries. As Robinson observes, "Mutual confidence is all that sustains the regime," and mutual confidence "may be damaged by mistaken suppositions of noncompliance as well as by correct ones."[11] The effectiveness of the regime is weakened by the absence of some politically neutral and credible means of sorting out whether and when violations have occurred. Repeated accusations between adversaries ultimately erode the norm.[12]

The remedy lies in international cooperation, which defines Robinson's second "basket" and involves measures rendering the weapon (again, for example, chemical or biological weapons) illegal, and at the same time providing some means of knowing whether or not others are abiding by their obligations. While an agreement with other states can give a government cause to change its intentions, a prudent government will change its behavior in terms of military capabilities maintained only with effective verification that others are complying with the agreement.

So what is "effective verification" and how does it get established? Robinson defines "sufficiency" in terms of detecting violations in time to take remedial steps, but perhaps the most succinct description has been provided by Wolfgang Panofsky in describing U.S. policy concerning the Threshold Test Ban Treaty:

> The terms "adequate" or "effective" require as a minimum that: 1) no violation that could endanger national security should remain undetected and unidentified; 2) a violation should be identified in sufficient time to allow remedial action to protect national security; and 3) no violation that interferes in a basic way with the essential purposes of the treaty should remain undetected and unidentified. I specifically emphasize that the impact of a possible violation on national security should be an essential element of any practical verification standard.[13]

Negotiating procedures that meet these criteria is always a difficult undertaking. To decide to negotiate an arms control or disarmament agreement requires balancing the value of an international cooperation "basket" against the value of a unilateral action "basket" (that is, maintaining the military capability or weapons system in question and going it alone). The characteristics of the unilateral action basket are well known to the government and may include significant costs to maintain weapons and capabilities. On the other hand, the international cooperation, or regime undertakings basket, is unknown and at best loosely defined when the decision to enter negotiations is made. It is only rather late in the negotiating process that the "barter" or balance between these baskets can really be evaluated.

A question critical to that balance is good faith. The intentions of other participants and the permanence of those intentions can be gauged only over the course of negotiations. For this reason and because committing to the regime is to foreclose an option (in many cases, to eliminate an existing capability), "the matter of establishing confidence in compliance with the regime lies at the very heart of the negotiations."[14] However, the matter of assurances cuts two ways. Just as it is necessary to be assured that other states will comply with the regime, it is also necessary that other states are assured of your continued good intentions and compliance with the regime. If the regime is to be of value, you must important to have some means of persuading other participants that you are still abiding by your obligations.[15] In the end, each state participating in the negotiations must choose between joining the new treaty and remaining outside it, between "national measures of self-reliance that afford less-than-perfect security against the . . . threat and collaboration in a less-than-perfect international regime."[16]

A somewhat different formulation of the international cooperation basket has been offered by Lewis Dunn,[17] who notes that the character of verification deemed necessary for an acceptable agreement can be described in terms of decisions made historically concerning the appropriate balance along a series of dichotomies:

- Trust and honor or suspicion and independent confirmation of other parties' compliance with their agreements
- Reliance on national intelligence monitoring or formal verification means to confirm compliance
- A more cooperative or a more adversarial approach to the verification process, including whether to emphasize that the purpose of verification is to provide assurance of compliance or to detect non-compliance
- The military or political significance of possible violations in designing verification regimes
- The potential gains of intrusive inspections in other countries as opposed to the risks of reciprocal inspections at home
- Pursuit only of strictly verifiable arms control limits or also of less verifiable but possibly politically valuable agreements.[18]

In sum, the political dynamics of formulating an arms control or disarmament agreement involve, for each state participating in the process, an evolution of thinking and analysis concerning whether the course ("basket") of self-reliance or the course ("basket") of international cooperation offers the better overall prospect of increased security and decreased uncertainty. During the process of negotiating an agreement, each state formulates positions concerning the level of verification it judges necessary for it to join (and acceptable to it in terms of intrusiveness at home). As negotiations proceed, a consensus evolves concerning what is necessary to make the security benefits outweigh the uncertainties and risk of exposure to the consequences of violations by others. Included in this process must be a realistic assessment of the technical practicalities; some kinds of verification may be highly desirable, but beyond existing (or foreseeable) technical capabilities. In addition, there is the uncertainty as to whether all these technical factors have been correctly evaluated. This uncertainty influences how the more direct uncertainties (compliance of other parties, and their intentions to continue to do so) are judged. In the end, some of the most active participants in the negotiations may decide that the new treaty does not adequately satisfy their security concerns (for example, India and China chose not to join the NPT, although India in particular had been an active participant in the negotiations).[19]

How Do States Verify?

The following chapter examines how the verification question has been addressed in the three treaty regimes dealing with different weapons of mass destruction. Historically, the first nonproliferation regime to take shape was that concerning nuclear weapons, the International Atomic Energy Agency safeguards system and the Treaty on the Non-Proliferation of Nuclear Weapons. By the end of the 1960s, when the NPT was being concluded, rapid progress was also made on the Biological Weapons Convention. Negotiation of the Chemical Weapons Convention, like the NPT, took a long time to conclude and did not open for ratification until the early 1990s. Each of these treaties (and its associated regime) shares some important characteristics. Each is intended to be universal, ultimately to be joined by all states in the international community. Each bans the possession (with the NPT exception for nuclear weapons states) of a weapon of mass destruction. And in banning a weapons technology, each treaty at the same time contemplates that states party to the treaty will maintain civilian facilities engaging in commercial and scientific activities technically close to and easily convertible to the production of the banned weapon.

This study investigates the degree to which these fundamental similarities led to the same approach to verification, and the degree to which different verification approaches were chosen. It also examines the factors that created the differences as well as the similarities.

If the model of verification and Robinson's model are accurate, we will see each reflected in the evolution of each of the treaties. However, if the security issues leading to nonproliferation treaties differ in some important respects from the security issues leading to treaties limiting the size of conventional military forces, either the model for verification (as opposed to the larger issue of regimes) will not apply, or Robinson's model of two "baskets" will fail in explanatory power.

The same question can be put a different way. To what degree are the verification measures included in nonproliferation and arms control treaties a function of:

- Some fundamental and intrinsic structure or logic of the verification problem
- Specific technical characteristics of the weapons to be banned or controlled and of related civilian activities that remain permitted, if any
- Random historical forces bearing no discernible relationship to either of the foregoing hypotheses.

Notes

1. Verification has become an element of multilateral agreements addressing a rather wide range of issues, not just issues dealing directly with limiting or eliminating kinds of weapons or other military security issues. See, for example, Philippe Sands, "Enforcing Environment Security: The Challenges of Compliance with International Obiligations," *Journal of International Affairs* 46, no. 2 (Winter 1993): 367-390.

2. A caveat is in order: the term "weapons of mass destruction" has become popular to describe nuclear, chemical, and biological weapons collectively. As a collective noun for these non-conventional weapons, it is useful. At the same time, each of these weapons is different in important ways from each of the others for reasons having to do with the technical characteristics of that weapon. These differences have been described well by James F. Leonard: "By their nature, chemical arms have a relatively limited range; they create regional rather than global security problems. In this, they are militarily more akin to conventional arms than to nuclear or biological weapons, even though nuclear, biological, and chemical weapons are generally classed together as 'weapons of mass destruction.' " In "Rolling Back Chemical Proliferation," *Arms Control Today* 22, no. 8 (October 1992): 13. As will become somewhat apparent in subsequent chapters, the term "weapons of mass destruction" conceals more than it illuminates.

3. The larger political significance of the fact that all five permanent members of the United Nations Security Council are also the five states recognized by the NPT (Treaty on the Non-Proliferation of Nuclear Weapons) as nuclear weapons-states cannot be overlooked, and is not overlooked by many other states in the international community.

4. The interaction between treaties and international norms of behavior is an interesting one. Treaties are sometimes negotiated to reflect pre-existing norms of behavior, as the Geneva Protocol claims to do with respect to nonuse of chemical weapons. But treaties also create new norms of international behavior, and certainly the Geneva Protocol sought, by once again formally declaring a proscription against using chemical weapons, to reinforce the very norm it claimed to reflect. After 1925 European powers did not use chemical weapons against each other, but continued to use them on occasion against non-Europeans; separating the effects of deterrence from normative factors with respect to other Europeans and nonwhites is clearly more complicated an endeavor than can be pursued here.

5. There are actually two international nuclear export control mechanisms, for purely historical reasons. One, the Nuclear Suppliers Group, included states conforming to the norm, which means France and many NPT parties. The other, the Zangger Committee, is restricted to NPT parties.

6. As one might expect, this is a topic on which senior U.S. officials are rarely direct. See, for example, an interview with "Ambassador Stephen J. Ledogar: The End of the Negotiations," in *Arms Control Today* 22, no. 8 (October 1992): 10. Also noteworthy is the new importance international organizations place on national intelligence, and their ability to receive, and act on, information provided by member states. See, for example, Rolf Ekeus, "The Iraqi Experience and the Future of Nuclear Proliferation," *The Washington Quarterly* 15, no. 4 (Autumn, 1992): 69 ff; and Maurizio Zifferero, "The IAEA: Neutralizing Iraq's Nuclear Weapons Potential," *Arms Control Today*, volume 23, number 3 (April 1993), 9-10.

7. These issues have spawned a vast and highly technical literature in the nuclear safeguards field, among others. See, for example, U.S. Nuclear Regulatory Commission, *Handbook of Nuclear Safeguards Measurement Methods*, ed. Donald R. Rogers, NUREG/CR-2078; also see the *Journal of the Institute for Nuclear Materials Management*.

8. Which is to say that, at the political level, some states do claim that verification (as, for example, by International Atomic Energy Agency safeguards inspections) do constitute proof. The whole point of this section is that deductive (absolute) proof is impossible, all verification operates on a scale of uncertainty.

9. Julian Perry Robinson, *Chemical Warfare Arms Control: A fFamework for Considering Policy Alternatives*, SIPRI Chemical & Biological Warfare Studies Number 2 (Stockholm: Stockholm International Peace Research Institute, 1985).

10. "Neighbors" is of course a relative term. For the United States (and perhaps a few other countries) national interests are so broad and extended (as are the ability of to protect those interests by projecting power globally) that important "neighbors" are global rather than regional.

11. Robinson, 6.

12. It seems that even so obvious a generalization as this one has its exceptions. Charles Van Doren has pointed out to the author that during the early nineteenth century the United States and Canada were constantly accusing each other of violating the Rush-Bagot Treaty regulating naval forces on the Great Lakes, without any significant detriment to the long-term relationship of trust between the two countries.

13. Wolfgang K. H. Panofsky, "Paths to a Test Ban Treaty: Two Views—Straight to a CTB," *Arms Control Today* 20, no. 7 (November 1990): 3.

14. Robinson, chapter 2, "The scope of the projected Chemical Weapons Convention," especially 22-23; quotation from 23.

15. Robinson, 72; see also Lewis Dunn, "Arms Control Verification After the Cold War," in *Arms Control Verification: Looking Back and Looking Ahead*, Richard C. Davis, Lewis A. Dunn, Sidney Graybeal, Ralph Hallenbeck, Patricia McFate, and Timothy Pounds (Washington, DC: SAIC, U.S. Department of Energy, June 22, 1993).

16. Robinson, 56.

17. Dunn.

18. Ibid., 9-10.

19. One case not explored in this analysis is that of the state which joins a treaty with the specific intent of using adherence to help conceal those very activities prohibited by the treaty. This analysis sought to focus primarily on the process of negotiating an agreement, as that is when the question of including or not including verifications measures is addressed. Situations in which a state actively participated in negotiating an arms control or disarmament agreement, including verification measures, with perfidious intent are relatively few and complex—but important. Such a state could be expected to oppose any verification measures, or at least have specific plans for how to evade those measures it accepted in the new treaty.

2. ❖ Controlling Nuclear Weapons

Early Evolution

The first modern regime to include specific measures to provide regular and ongoing verification of compliance was that directed at the problem of nuclear weapons. This regime, which had its origins in the final days of World War II, represents a series of efforts, first for global disarmament, then for quite limited controls, and since the mid-1960s toward a more effective regime. It is this later development, the Treaty on the Non-Proliferation of Nuclear Weapons (NPT), on which we shall concentrate, as it is only with the NPT that there was a universal (open to all states) treaty prohibiting the further proliferation of nuclear weapons and providing for verification of that prohibition. But it is important to know how the nuclear nonproliferation regime and its verification system developed.

The nuclear nonproliferation regime developed in several stages. The first efforts took place in the first months and years after World War II. President Eisenhower's "Atoms for Peace" proposal initiated the second stage, which included creation of the International Atomic Energy Agency (IAEA). A third stage began in the 1960s with negotiation of the NPT. As we shall see, the nature of verification in the nuclear nonproliferation regime was influenced as much by historical contingencies as by the substantive nature of the problem, perhaps more so.

In November 1945, just 3 months after the United States dropped the first atomic bombs on Hiroshima and Nagasaki, the president met with the prime ministers of Great Britain and Canada to discuss post-war security issues and the atomic bomb. That meeting produced a joint policy statement, expressing the view that:

- There can be no monopoly on nuclear weapons, and no effective defense against them

- Nuclear energy can be a source of great benefits to mankind
- It is vitally important to prevent proliferation and nuclear war and to pursue the peaceful benefits of nuclear energy
- This is the responsibility of the international community, not just a few nations.

In words to become famous (albeit frequently taken out of context), the trilateral declaration judged, "No system of safeguards that can be devised will of itself provide an effective guarantee against the production of atomic weapons bent on aggression."[1] The statement did not elaborate on what additional measures would be necessary; that would come later with the Baruch Plan, but it did note,

> We are not convinced that the spreading of the specialized information regarding the practical application of atomic energy, before it is possible to devise effective, reciprocal, and enforceable safeguards acceptable to all nations, would contribute to a constructive solution of the problem of the atomic bomb. On the contrary, we think it might have the opposite effect.[2]

In other words, nuclear energy had important peaceful applications, but the technology to pursue these must wait for effective measures to prevent the spread of the atomic bomb. (This is in strong contrast to the subsequent course of events, much less the situation with biological and chemical weapons.)

The very first action of the new United Nations (U.N) General Assembly when it met in January 1946 was to create a U.N. Atomic Energy Commission (UNAEC) to investigate steps concerning "exchange of information, control to ensure only peaceful use of atomic energy, elimination of atomic weapons and other weapons of mass destruction, and effective safeguards."[3] In establishing the UNAEC, the General Assembly noted an important concept: the policy aim that "the fruits of scientific research should be freely available to all nations" was coupled with the pursuit of arms control and disarmament. The latter objective entailed "effective safeguards by way of inspections and other means to protect complying States against the hazards of violations and evasions." Both the purpose (to "protect complying States") and the methods ("inspections and other means") of safeguards were clearly articulated as early as 1946.[4]

What remained undefined was the larger architecture of which safeguards (verification) would be one part.

The United States presented a proposal for this architecture, which originated in the Acheson-Lilienthal Report but which in its presentation to the UNAEC had become the significantly different Baruch Plan.[5] It represented a dramatic step in governmental thinking about how to deal with international security issues—virtually the establishment of an international government in the area of nuclear energy: "The control and development of atomic energy must be international and should be entrusted to an agency which for present purposes is called the atomic development authority."[6] Under the U.S. plan, "the national and private possession, manufacture, and use of atomic weapons shall be outlawed," and the development authority would have "managerial control or ownership of all atomic energy activities potentially dangerous to world security" as well as responsibility "to control, inspect, and license all other atomic energy activities; to engage in atomic energy research and development; and to assure that the benefits derived from such research and development shall be available to the peoples of all the signatory States."[7] Clearly this was an innovative and radical proposal; and it quickly proved to be too much so.

The Soviet Union rejected the proposal for several reasons.

- The USSR would not accept the sequence of first establishing a control regime to which the USSR and other countries would be subject, with nuclear disarmament through the elimination of U.S. weapons only coming afterwards.
- The USSR would not accept amending U.N. Security Council procedures with respect to nuclear non-proliferation issues to eliminate the veto rights of the members with permanent seats (a theme to be replayed later with respect to the Biological Weapons Convention).
- The USSR considered U.S. ideas as too intrusive and violating sovereignty[8] although it agreed with the need for a control regime.

Within 2 years of its establishment, the UNAEC was moribund, a victim of both the Cold War and the difficult legal and political issues it sought to address. The impasse remained for 5 years, until President Eisenhower made his famous "Atoms for Peace" speech before the U. N. General Assembly in December 1953. While his speech dealt with both the control and promotion of nuclear energy, Eisenhower chose to emphasize the opportunities to be realized rather than the dangers to be avoided. This decision, made personally by the president, was intended to "develop a peaceful alternative to nuclear weapons" as a way to "break the deadlock in disarmament negotiations with the Soviet Union."[9] Eisenhower's choice has shaped the course of nuclear nonproliferation efforts ever since. His approach was immediately successful, but placing development of peaceful uses ahead of effective international controls would prove fateful for the verification system in the longer run.

In Atoms for Peace, the United States proposed establishment of a new international organization specifically to pursue peaceful uses of nuclear energy. The organization would receive contributions of nuclear material from those countries owning stockpiles and then distribute the natural and enriched uranium to member states for use in producing electricity and exploring scientific, agricultural, industrial, and medical applications of radioactive materials. This organization would also provide technical information on peaceful uses of nuclear energy. In return, a country receiving materials or information would accept verification by the organization to assure that any technology or nuclear materials received were being used only for peaceful purposes.

Negotiations on an international atomic energy agency were conducted from June 1954 through October 1956. At first the United States and the Soviet Union talked bilaterally. The United States then consulted with allies, but in response to criticism of this approach widened the group to include the Soviet Union, Czechoslovakia, Brazil, and India. Negotiations on the Statute, or charter, of the International Atomic Energy Agency (IAEA) were completed in a conference open to all members of the United Nations.[10] The original discussions in the UNAEC had focused on international machinery to control nuclear materials and technology to verify the prohibition on nuclear weapons; dissemination of scientific and technical information was a distinctly secondary function. The Atoms for Peace proposal reversed this, giving dissemination of information and pursuit of civil applications the higher priority, albeit as a means of achieving consensus on a system which did both.

A key aspect of the negotiations on the new agency dealt with the appropriate balance between its two central functions and whether the issue was nuclear disarmament, control of uses, or simply verification of member state undertakings. The USSR at first maintained its previous stance that the objective was nuclear disarmament but argued against intrusive safeguards; India took this view as well. For the United States, the objective was control of peaceful uses and arms control, and the United States supported strong safeguards. A number of Western European and less-developed countries were actively pursuing nuclear research; for them, extensive and intrusive controls were not acceptable.

Resolution of this issue was key to the success of the Statute Conference and the organization it was to create. The relative emphasis on promoting peaceful uses and controlling proscribed uses was also central to the status of verification procedures (safeguards) in the system. A system that emphasized promotion would not emphasize strong and intrusive safeguards to the same degree as a regime in which controlling proliferation had the highest priority.

The Internation Atomic Energy Agency Statute and Early Safeguards

The compromise achieved is reflected in Article II of the Statute of the IAEA, which identifies two objectives for the organization. First, "to accelerate and enlarge the contribution of atomic energy to peace, health and prosperity throughout the world." Second, to "ensure, so far as it is able, that assistance provided by it or at its request or under its supervision or control is not used in such a way as to further any military purpose."[11] To accomplish this second objective, one function of the Agency is to:

> establish and administer safeguards designed to ensure that special fissionable and other materials, services, equipment, facilities, and information made available by the Agency, or at its request or under its supervision or control are not used to further any military purpose; and to apply safeguards, at the request of the parties, to operations under any bilateral or multilateral arrangement, or, at the

request of a State, to any of that State's activities in the field of atomic energy.[12]

Eisenhower's Atoms for Peace speech contemplated a regime in which the IAEA would be the conduit for international cooperation and commerce in nuclear material and technology and the Statute reflects this. According to the Statute, safeguards are to apply even to services, equipment, and information.

But Atoms for Peace quickly took on a second face even as the Statute was being negotiated. Atoms for Peace also produced the Atomic Energy Act of 1954, which authorized supply of research laboratories and reactors (and highly enriched uranium fuel) directly to other countries. This bilateral cooperation was implemented by formal "agreements for cooperation" between the United States and the receiving government, stipulating conditions of the assistance. One condition was safeguards, verification that exported commodities were being used only for peaceful purposes. Inspections were performed by U.S. authorities, with the understanding that once the international organization envisioned in Eisenhower's speech was established, it would take over safeguards. Similar bilateral cooperation programs were quickly established by Britain and the USSR, and later France and Canada. The resulting fabric of bilateral arrangements (cooperation and obligations) essentially undermined the original premise of the Statute. No longer was the IAEA the sole conduit for knowledge, technology, and materials. There was no obligation to foreswear nuclear weapons, only to separate safeguarded activities from any nuclear weapons related activities. Safeguards on technology[13] would not be attempted for 20 years, long after training and technology had permitted the creation of many indigenous programs free from any safeguards or IAEA oversight.

Because the bilateral cooperation system was already well developed by 1956, the negotiators of the Statute recognized the need for the Statute to address this situation. Hence the second part of the provision quoted above, permitting the IAEA to apply safeguards, when requested, to items obtained bilaterally or developed indigenously. In the long term this provision would prove to be the more important; it was this latter clause that permitted the IAEA to become the verification agent for the NPT.

Although the IAEA was established as an independent international organization and not as a specialized agency of the United Nations operating under the Charter, the Statute explicitly subordinates the IAEA to the United Nations, and especially the Security Council, "as the organ bearing the main responsibility for the maintenance of international peace and security."[14]

The IAEA General Conference "may discuss any questions or any matters within the scope of this Statute . . . and may make recommendations to the membership of the Agency or to the Board of Governors or to both on any such questions or matters."[15] It may also suspend a member "from the privileges and rights of membership"[16] if that member State has "persistently violated the provisions of this Statute or of any [safeguards, project, or supply] agreement entered into by it pursuant to this Statute."[17]

The IAEA Board of Governors has very broad powers "to carry out the functions of the Agency in accordance with this Statute, subject to its responsibilities to the General Conference."[18] In practice, the Board approves all agreements between the Agency and member states, such as project and supply agreements (which, in addition to transferring a nuclear facility, nuclear material, or equipment, also require safeguards on the supplied items) and safeguards agreements. In practice, it is the Board, and not the General Conference, that oversees the Secretariat's work in implementing safeguards, and it is also the Board that receives reports from the Director General on safeguards matters. If the Director General informs the Board that a safeguards agreement has been violated, the Board

> shall call upon the recipient State to remedy forthwith any non-compliance which it finds to have occurred. The Board shall report the non-compliance to all members and to the Security Council and General Assembly of the United Nations. In the event of failure of the recipient State to take fully corrective action within a reasonable time, the Board may take one or both of the following measures: direct curtailment or suspension of assistance being provided by the Agency or by a member, and call for the return of materials and equipment made available to the recipient member. The Agency may also, . . . suspend any non-complying member from the exercise of the privileges and rights of membership.[19]

While these sanctions fall far short of the acts of force necessary to wrest a nuclear weapons program from a renegade state, they are far reaching and (until the Chemical Weapons Convention comes into force) effectively exceeded only by the rights afforded to the Security Council itself.

The Board of Governors has reported several instances of noncompliance to the Security Council. The first involved Iraq, based on discoveries subsequent to Iraq's defeat in the Gulf War. The second involved Romania, which itself reported violations by the previous Communist government. The most significant case, because the Agency itself detected the noncompliance (in Iraq noncompliance was discovered as a byproduct of defeat in war, and Romania reported its noncompliance under the previous Communist government), occurred in May 1993 when the Board informed the Security Council that the IAEA was unable to perform its verification obligations in North Korea and suspended all IAEA technical assistance to North Korea.[20] In the case of Iraq, the IAEA has actually destroyed nuclear facilities and removed nuclear material, but these actions were taken under Security Council mandate, not the Statute.

The IAEA has also reported instances in which it could not perform its safeguards responsibilities, for example with respect to two unidentified states (widely understood to be Pakistan and India) in 1983. While not, strictly or legally speaking, a report of noncompliance, when the Secretariat reports to the Board (or the IAEA reports to the United Nations) that the inspectorate is unable to fulfill its verification responsibilities because of actions by a member state, this is one step short of reporting actual noncompliance and an effective lever to obtain compliance with safeguards obligations.

The Secretariat (the staff of the IAEA) is headed by the Director General and is charged to "fulfill the objectives and functions of the Agency." In this capacity all personnel are to be truly international civil servants:

> In the performance of their duties, the Director General and the staff shall not seek or receive instructions from any source external to the Agency. They shall refrain from any action which might reflect on

their position as officials of the Agency; subject to their responsibilities to the Agency, they shall not disclose any industrial secret or other confidential information coming to their knowledge by reason of their official duties for the Agency.[21]

The powers of the IAEA with respect to safeguards are listed in Article XII of the Statute. As originally conceived, the IAEA was to have significant powers. But because subsequent developments did not take the intended course, some of the language detailing Agency rights and obligations must be interpreted carefully. The Statute gives the Agency authority to:

- Examine the design of specialized equipment and facilities, to approve it only from the view-point of assuring that it will not further any military purpose, . . and that it permit effective application of safeguards
- Require the maintenance and production of operating records to assist in ensuring accountability for source and special fissionable material
- Approve the means to be used for the chemical processing of irradiated materials solely to ensure that this chemical processing will not lend itself to diversion of materials for military purposes
- Require that nuclear materials . . . produced as a by-product of peaceful activities under safeguards shall also remain subject to safeguards;
- Require deposit with the Agency of any excess fissionable materials produced or recovered as a by-product . . . to prevent stock-piling of materials not being utilized directly in peaceful activities
- Send inspectors . . . designated by the Agency after consultation . . . with the State . . . who shall have access at all times to all places and data and to any person who by reason of his occupation deals with materials, equipment, or facilities which are required by this Statute to be safeguarded, as necessary to account for source and special fissionable materials . . . and to determine whether there is compliance with the undertaking against use in furtherance of any military purpose.[22]

Clearly the safeguards provisions of the Statute still resonated with ideas originally put forth in the Baruch Plan. But now the powers of the inspectorate were extended not only to data collection but also to judging compliance:

> The staff of inspectors shall also have the responsibility . . . of determining whether there is compliance with the undertakings . . . and with all other conditions of the project prescribed in the agreement between the Agency and the State. . . . The inspectors shall report any non-compliance to the Director General who shall thereupon transmit the report to the Board of Governors.[23]

A number of aspects of this paragraph warrant comment. First, it is the inspectors, and not their supervisors or the Director General, who are to determine compliance ("the inspectors shall report any non-compliance"). The Director General appears to have only procedural functions in this situation—reporting the finding to the Board. However, this same paragraph in the Statute appears to give the Board the power to determine compliance: "The Board shall call upon the recipient State or States to remedy forthwith any non-compliance *which it finds to have occurred.*"[24] In recent practice the Director General reports to the Board that the Secretariat is unable to confirm compliance, along with the reasons for this inability, and the Board makes the formal determination.

One fundamental aspect of the IAEA safeguards process is that the IAEA must negotiate a "safeguards agreement" with the state (or states) to confirm the obligation to provide safeguards and to acquire the rights necessary to fulfill this obligation. The Statute requires a safeguards agreement only in the original context of being the conduit through which a state received technology and material.[25] There is no requirement in the Statute to negotiate a safeguards agreement when a state voluntarily submits indigenous facilities or materials, or an exporting state requires safeguards.[26]

The Statute established a system including onsite inspections along with considerable data flow to the inspectorate for evaluation and cross-checking. But as originally conceived, the inspectorate also had authority to ensure that nuclear facilities are designed and built to facilitate safeguards and consistent with peaceful nuclear activities but not activities prohibited by the Statute (activities identified only as "any military purpose"—a vague but useful characterization that will not be overtaken by technological developments). A system in which these measures were fully implemented would not only have substantial capability to verify that individual member states were complying with their obligations but would also afford the international organization a significant degree of control over those activities.

The Agency's first efforts to establish such a safeguards system were halting and difficult. The ad hoc nature of the arrangements first proposed were attacked by the very states which "opposed the appointment of staff to study in advance the problems of safeguards."[27] The first standardized system was adopted in 1962. Experience with this system rather quickly led to its revision and extension. A second system was initially promulgated in 1965 and quickly amended to cover several additional types of nuclear facilities.[28] This document, INFCIRC/66/Rev.2, describes principles more than specifies practice, stipulating "a list of rules to govern the application of safeguards as well as the general approaches and procedures to be used."[29] Portions of this document are included, and others incorporated by reference, in each safeguards agreement. Ironically, it was objections to this safeguards system, which was only defined between 1965 and 1968, which led to the significantly different safeguards approach under the NPT, even though negotiations on the NPT were completed and the document opened for signature in 1967.

Beginning in the early 1970s, several steps were taken to strengthen or clarify safeguards under INFCIRC/66. Responding to an Argentine effort to limit safeguards on an imported reactor to 5 years, in 1973 the Board established a formal policy on duration and termination of INFCIRC/66-type safeguards, requiring that safeguards continue until the item is no longer usable for nuclear purposes.[30] The next year the Board accepted the Director General's interpretation of "any military purpose" to include any nuclear explosive device, as nuclear explosives were not readily distinguishable based on intended use.[31] At about this time the Agency finally addressed safeguards on technical information, introducing into safeguards agreements covering facilities supplied by one country to another the requirement that any subsequent facility using the same technology would also be subject to safeguards under the same agreement.[32]

As noted above, INFCIRC/66 safeguards rights and responsibilities apply only to materials and facilities transferred through the IAEA, or more frequently, made subject to IAEA safeguards as a result of bilateral agreement between the supplier and recipient. As such, IAEA safeguards could offer no judgment regarding indigenous nuclear activities (unless voluntarily submitted to safeguards) and specifically could not judge whether the same state had a covert nuclear weapons program in addition to its safeguarded peaceful nuclear program. Recognizing this defect, in 1961 Ireland introduced a resolution in the General Assembly calling for an international agreement banning transfer or acquisition of nuclear weapons.

The Non-Proliferation Treaty

For several years the proposed agreement was discussed in the Eighteen Nation Disarmament Committee (ENDC) with little progress. The United States and Soviet Union agreed that the treaty must prohibit nonnuclear-weapon states from manufacturing or otherwise acquiring nuclear weapons, and prohibit nuclear-weapon states from providing nuclear weapons to nonnuclear-weapon states, but argued over the implications of these points for NATO.[33] Many nonnuclear-weapon states emphasized other themes: a need for the nuclear-weapon states to pursue arms limitations; the application of nuclear explosives technology to peaceful uses—Peaceful Nuclear Explosives (PNEs); access by the nonnuclear-weapon states to scientific and technical information and technology relevant to peaceful purposes; and security assurances for non-nuclear weapon states.[34]

Finally, after bilateral talks, in August 1967 the U.S. and Soviet co-chairmen tabled separate but identical drafts of a convention. This text was discussed first in the ENDC and then in the U.N. General Assembly, and on June 12, 1968, the U.N. General Assembly adopted a resolution commending the text and encouraging adherence. On July 1, 1968, the Treaty was signed by over 60 nations, and on March 5, 1970, the NPT entered into force.

The first significant feature of the NPT as an arms control agreement is that it divides all potential parties in two groups: those that already possessed and had detonated nuclear weapons prior to January 1, 1967, and those that had not.[35] Nuclear weapon states commit not to assist "any recipient whatsoever" in acquiring nuclear weapons or control over any nuclear explosive devices.[36] Nonnuclear weapon states commit not to "receive the transfer . . . or control over" nuclear weapons or nuclear explosives, and "not to manufacture or otherwise acquire nuclear weapons or other nuclear explosive devices; and not to seek or receive any assistance " in their manufacture.[37] To confirm (in part) these obligations, each nonnuclear weapon state:

> undertakes to accept safeguards, as set forth in an agreement to be negotiated and concluded with the International Atomic Energy Agency in accordance with the Statute of [the IAEA] and the Agency's safeguards system, for the exclusive purpose of verification of the fulfillment of its obligations assumed under this Treaty with a view to preventing diversion of nuclear energy [sic] from peaceful purposes to nuclear weapons or other nuclear explosive devices.[38]

This is in contrast to previous safeguards obligations, which had precluded any military use but did not prohibit manufacture or possession of PNEs. It is important to note, in this rather belabored language that the "exclusive purpose of verification" is to confirm the "fulfillment of its obligations assumed under this Treaty." Because only nonnuclear weapon states are required to accept safeguards, as the new NPT system was shaped and implemented, these states insisted that inspections were to verify national declarations of nuclear material, not to search the whole country for evidence of violations. Many perceived safeguards to be more confidence-building measures than rigorous verification. (Ironically when, after the 1991 Gulf War, the safeguards system is found wanting, this language in the NPT will provide the legal basis for implementing corrective measures extending safeguards to include searching for undeclared activities which violate NPT and safeguards undertakings.)

While Statute safeguards apply to nuclear facilities, other specified materials and equipment, and even technology, the NPT stipulates that safeguards shall be applied only to "source and special fissionable material whether it is being produced, processed or used in any principal nuclear facility or is outside any such facility," but "shall be applied to all source or special fissionable material in all peaceful nuclear activities within the territory of such State, under its jurisdiction, or carried out under its control anywhere."[39] Many countries objected to the intrusiveness inherent in applying safeguards to facilities and equipment and argued that applying safeguards only to nuclear material would be adequate, because safeguards would apply to all nuclear material. Thus the ambit of NPT safeguards is, on the one hand more limited than under the original Statute system, but in another respect much more comprehensive. Clearly the authors of the NPT contemplated a new safeguards system.

A further limitation is that safeguards will apply only to nuclear material in "peaceful nuclear activities."[40] The NPT does not define "peaceful nuclear activities" or characterize nonpeaceful nuclear activities, nor does it ever address, or even refer directly to, the question of which nonpeaceful uses of nuclear energy (or nuclear material) are not proscribed.[41] This becomes an important issue in the subsequent development of the NPT safeguards regime.

Another reflection of nonnuclear weapons states' concerns about the intrusiveness of Statute safeguards is the requirement that NPT safeguards "shall be implemented in a manner designed to comply with Article IV of this Treaty [guaranteeing all parties access to the beneficial uses of nuclear energy], and to avoid hampering the economic or technological development of the Parties or international cooperation in the field of peaceful nuclear activities."[42] In addition, NPT safeguards are to be implemented in accordance with "the principle of safeguarding effectively the flow of source and special fissionable materials by use of instruments and other techniques at certain strategic points."[43] This "principle" of focusing on "certain strategic points" will prove very important in the history of NPT safeguards; it is the rock on which the system nearly foundered in Iraq.

Like safeguards under the Statute itself, the safeguards obligation under the NPT is not self-executing. Each nonnuclear-weapon state (or group of states) "shall conclude agreements with the [IAEA] to meet the requirements of this article . . . in accordance with the Statute."[44] While the NPT requires each party to conclude a safeguards agreement with the IAEA within 18 months of ratifying or acceding to the NPT, this has not always been done. Most states that fail to conclude a safeguards agreement in a reasonable time in fact do not have any significant nuclear activities to place under safeguards, and so the failure is of legal and diplomatic but not practical consequence. North Korea is a notable exception. The only recourse in such cases is for the Board of Governors (or the NPT Depositary governments) to refer the matter to the Security Council.[45]

Information Circular 153

In response to the NPT provisions relating to the establishment of a specific safeguards system for its implementation, in 1972 the IAEA's Board of Governors created a special committee of the Board, open to all members of the IAEA and known as the Safeguards Committee, to establish new safeguards procedures specifically tailored to the NPT. Over the course of some 8 months, the Safeguards Committee produced a model safeguards agreement, "The Structure and Content of Agreements Between the Agency and States Required in connection with the Treaty on the Non-Proliferation of Nuclear Weapons." After review and approval by the Board of Governors, this document was published as Information Circular (INFCIRC/) 153.

It establishes a two-tiered safeguards system.[46] The foundation is a national (or possibly multinational, as in the Euratom and Argentine/Brazilian cases)[47] system of accounting for and controlling nuclear material. This system maintains facility- and national-level records and provides reports to the IAEA, which in turn is responsible for performing the measures necessary to confirm the accuracy and completeness of the national system. This takes a substantial burden off the IAEA while permitting the individual state to tailor its national system to its legal and political situation. Burdens (intrusiveness) on the state and operator are in some respects reduced even as the IAEA is relieved of considerable detail work.

Part I of the document identifies the principal legal and administrative aspects of the safeguards agreement; most is an elaboration (and in many respects a narrowing) of the inspection rights IAEA was given under the Statute. Part I also addresses the

NPT provisions permitting parties to exclude from safeguards nuclear material being used in nonpeaceful activities that do not violate the NPT. In the event a state plans to use nuclear material in a "non-proscribed military activity not in conflict" with its NPT obligations, it must conclude arrangements such that the IAEA is informed of "the period or circumstances during which safeguards will not be applied."[48] By agreement when INFCIRC/153 was negotiated, safeguards must be applied while nuclear material is being processed for such use, and reapply as soon as the material is no longer in active use.[49] To date no state has requested Agency agreement to such arrangements.

Part II elaborates in greater detail the rights and obligations of the IAEA, the state, and individual facility operators. These arrangements can be summarized relatively succinctly, but perhaps of greatest significance are the three paragraphs under the topic of "Objective of Safeguards," the first of which stipulates:

> the objective of safeguards is the timely detection of diversion of significant quantities of nuclear material from peaceful nuclear activities to the manufacture of nuclear weapons or of other nuclear explosive devices or for purposes unknown, and deterrence of such diversion by the risk of early detection.[50]

This paragraph contains a number of important concepts and innovations. The terms "timely detection" and "significant quantity" indicate that safeguards are to have a specific quantitative basis, but the document itself does not establish the specific quantities, permitting variation according to technical circumstances or as technology and commercial practice evolve. While not explicit, it is clear that the "significant quantity" is to be related to some quantity that can lead to a militarily significant activity, the construction of a nuclear weapon. However, the paragraph also stipulates that the IAEA need only confirm that a "significant quantity" is missing; it is not required to demonstrate that the missing material has in fact been diverted to a prohibited use. This substantially eases the burden of proof on the inspector and leaves to the state the ultimate responsibility to demonstrate, when it cannot account for all nuclear material, that it has not violated its treaty obligations. Finally, this paragraph states that safeguards have two related objectives: detection of diversions in violation of obligations and deterrence of such violations. Safeguards are both to strengthen the resolve of states parties to abide by their obligations, and to catch them when they do not.

The next two paragraphs of this section state that materials accountancy is to be "a safeguards measure of fundamental importance, with containment and surveillance as important complementary measures" and that the "technical conclusion of the Agency's verification activities shall be a statement" regarding the accounting balance of nuclear material for each specified locale (which may be several facilities, a facility, or part of a facility, depending on circumstances).[51]

These two paragraphs have been the subject of much interpretation and debate since 1972. On the one hand, they have led to an elaborate system for precisely accounting for nuclear material, by the gram for plutonium and enriched uranium. This is a system that, with modern advances in technology, has been able to detect rather small fluctuations in the measured values of nuclear material in processes involving chemical and physical (solid to liquid to solid) transformations.But which is nonetheless frequently criticized for limitations on the size of detectable fluctuations as being too large. At the same time, this system has been blamed for a too great focus on accounting for declared nuclear material at the expense of a wider focus on whether all the data obtained during an inspection are consistent with full compliance, or whether there are signs of prohibited activities that do not appear in the accounting process.[52]

Under normal circumstances the IAEA is given access to "strategic points" in a facility, and these strategic points are to be identified by the Agency, based on design information and walking through the plant, in consultation with the state and facility operator. Until a "facility attachment" specifying these points and other aspects of "routine" inspection have been concluded, IAEA inspectors perform "ad hoc" inspections permitting interim access to all locations within the facility. This procedure for receiving and reviewing design information, and then identifying the specific points in the facility to which access is required for routine inspection purposes, is a major innovation in INFCIRC/153, in contrast to safeguards practice under the older Statute based system. Its purpose is to reduce the intrusiveness of inspections. In any regulatory process, commercial operators desire as much predictability and as little interference with the production process as possible. However, such desires are not entirely consistent with effective verification (as would be demonstrated by Iraq).

For this reason, in addition to the system for regular inspection of all locations containing declared nuclear material, INFCIRC/153 provides for "special inspections" either to verify information on abnormal situations reported by the state ("special reports") or

> If the Agency considers that information made available by the State, including explanations from the State and information obtained from routine inspections, is not adequate for the State to fulfill its responsibilities under the Agreement.[53]

In other words, whenever the IAEA finds itself unable to conclude that all nuclear material indicated in the state's accounting and its own inspection reports "remained in peaceful nuclear activities or was otherwise adequately accounted for,"[54] the IAEA may request a special inspection, which can include "access to information or locations in addition to the access specified in paragraph 76 for ad hoc and routine inspections."[55] This additional access (to documents or to "suspect sites") is to be determined in consultation with the inspected state, and any disagreement is to be handled through normal dispute resolution procedures (an arbitral panel, with appeal to the International Court of Justice), except in cases where the Board of Governors determines, based on a report from the Director General, that access "is urgent and essential in order to ensure verification that nuclear material subject to safeguards under the agreement is not diverted." In this case the state is obligated to permit access "without delay, irrespective of whether procedures for the settlement of a dispute have been invoked."

Once INFCIRC/153 was formally approved by the Board in 1972, the Secretariat faced the daunting task of negotiating new safeguards agreements with all those nonnuclear weapon states that had by that date ratified the NPT. As was the case with previous steps in the evolution of IAEA safeguards, no sooner was the deal agreed than those states opposing specific measures included in the final result began to push implementation back to the more limited approach espoused during negotiation of the new approach.[56] Because INFCIRC/153 contains the model safeguards agreement, the language of all agreements is virtually identical, reflecting only different options available for certain administrative issues. But the IAEA also had to negotiate more technical operational and procedural documents, known as Subsidiary Arrangements and Facility Attachments, which, because they contained detailed design facility information, are held in confidence between the Agency and the individual state. In negotiating these documents many states sought to regain positions lost during negotiation of INFCIRC/153.

This was especially the case with respect to frequency and scope of inspector access to facilities, but also with types of safeguards equipment and procedures the state or facility operator would accept. While INFCIRC/153 clearly establishes the Agency's right to conduct unannounced inspections, objections were frequently raised on both practical ground and "as a matter of principle,"[57] and the few instances in which unannounced inspections were conducted involved such unusual circumstances as to demonstrate that the right was not accepted in practice.[58] For all practical purposes, IAEA inspection rights became limited to declared facilities, and normally to previously agreed locations ("strategic points") within those facilities, following due notice and coordination with the government and facility operator.

Critics noted the significance of these limitations, some arguing that IAEA safeguards were no better than a confidence-building measure, and one not instilling much confidence at that. But until the 1991 Gulf War the political will among member states to respond to these critics did not exist; they were instead ignored.[59] Following the Gulf War the international community discovered that Iraq had been conducting an extensive covert nuclear weapons program, including several large facilities working on different uranium enrichment methods, and some very small work on plutonium production and handling conducted in part at a facility being inspected.[60]

These discoveries energized the international community and led to a number of actions in the Board of Governors. Between February 1992 and June 1993 the Board took a number of important steps.[61]

First, it reaffirmed the right of the IAEA to conduct special inspections, and specifically to request, and even to insist upon, access to sites not declared by the state as relevant to nuclear activities. This right to inspect undeclared sites is clearly articulated in INFCIRC/153, but had not previously been used (perhaps because of the belief of senior Secretariat officials that the Board would not support inspectors in the event of a confrontation[62] or because no situation providing a need and sufficient evidence to support such a request had ever arisen).[63]

Second, the Board clarified the question of what information may provide the basis for the Director General requesting a special inspection. Previous understanding had been that it must be some combination of information provided by the state and generated by inspectors in the course of inspections in that state. The Board explicitly rejected this limitation and affirmed that "non-safeguards" information, whether arising from Agency technical cooperation activities, public records, or news media, or provided to the Secretariat by another member state, could also be used. The Director General was authorized to use any information coming into his possession, provided that he was prepared to take responsibility for judging the information to be of sufficient credibility to warrant pursuit.

While national intelligence information was not explicitly a part of this arrangement, the debate makes it clear that the Governors understood the breadth of the language in the decision taken, and intended to include intelligence.[64] This understanding was confirmed in practice in February 1993, when the Secretariat presented to the Board extensive evidence that North Korea was not complying with its NPT safeguards obligations. The evidence included both analysis of samples taken during inspections and satellite imagery (obviously provided by the United States) of facilities the IAEA sought to inspect but from which inspectors had been barred by North Korea. The Board discussion then and subsequently focused on the issue of the noncompliance and not on the nature of the evidence used to demonstrate that noncompliance.

Third, the Board determined that language in all Subsidiary Arrangements should be amended to ensure that states are obligated to inform the inspectorate of new facilities, or significant design changes to existing facilities, before the facility is built or modified, and to ensure that complete "as built" information is available to inspectors six months before nuclear material is introduced.

Finally, the Board expanded significantly the types of information concerning exports and bilateral cooperation which states are to provide to the IAEA. The Agency also requested annual information from all member states (not just NPT parties) regarding domestic production of uranium ore concentrate. Unlike the preceding changes to IAEA practice, this last is to be implemented on a voluntary basis

What is most significant about these and other safeguards reforms now under consideration is that none required amendment of the Statute, the NPT, or even of INFCIRC/153 and safeguards agreements negotiated pursuant to it. The legal basis was always there, it only needed a different interpretation. The change in interpretation required political will among member states, and until the revelations about Iraq's multiple safeguards violations and extensive nuclear weapons program, that political will had been lacking. Perhaps the most important question is how long this political will can last, and whether the principal members of the Board and senior officials in the Secretariat can resist gradual erosion of these gains as the threat fades in immediacy.

Analysis

The IAEA safeguards system is very different from other systems. Perhaps because of its heritage in the Baruch Plan, the IAEA safeguards system is the most completely international of all treaty verification systems. The IAEA's Secretariat is responsible for performing all elements of the verification function; that is, it receives reports, conducts inspections and collects other information, and then evaluates that information and forms a judgment regarding compliance or noncompliance based on the information available to it. Sometimes the judgment is instead that the Agency cannot determine compliance (because of inadequate information or access).

Of great significance is the fact that the IAEA does not verify compliance with the NPT, but verifies compliance with a safeguards agreement between the state and the IAEA, which the state entered into to fulfill an obligation under the NPT. The IAEA does not, even by the current interpretation of its safeguards role, seek to detect activities unrelated to the diversion or undeclared possession of nuclear materials. Neither the IAEA nor its member states have ever formally addressed the question of whether the IAEA should expand verification to include detection of activities related to nuclear weapons (such as fuzing and firing systems, high-explosives testing, or delivery system development) but not involving nuclear material. Nor does the IAEA verify any of the obligations nuclear weapon states undertake in Article 1 of the NPT. The original objective in 1945 was global nuclear disarmament. The NPT is a pragmatic effort to stop the proliferation of nuclear weapon states but recognizes that global nuclear disarmament was not, and may not be, achievable.[65]

That same sense of pragmatism is reflected in the verification system for the NPT. When the IAEA finds a state in noncompliance, as it has done on several occasions (Romania, Iraq, and North Korea), that noncompliance is with an IAEA safeguards agreement. It is for NPT parties and the Security Council to judge whether or not a state has violated its NPT obligations. In fact, there is no body specifically assigned responsibility for determining whether or not a state is in compliance with the NPT. This odd and partial relationship between the IAEA and the NPT arises because the IAEA was created first, and then used as the vehicle for verifying NPT undertakings. Had all the significant members of the IAEA become parties to the NPT early on, the partial relationship might not have developed. But as long as France, Spain, China, India, Pakistan, Israel, South Africa, Argentina, and Brazil were all outside the NPT (which was the case until recently, and half these states still remain outside the NPT), there would be strong objections to making the IAEA directly responsible for overseeing the NPT as such.

In the case of nuclear safeguards, it appears that the course of verification efforts was defined long before the IAEA system was established and long before there was a civil nuclear industry on which to apply such safeguards. The United States (in fact, a few wise and prescient individuals serving on a senior advisory committee) identified the broad themes of the nonproliferation system and its verification elements, and the subsequent course of events has essentially been a gradual realization of some (but decidedly not all) of those ideas.

At present the international nuclear safeguards system is largely consistent with that envisaged by those who negotiated the NPT safeguards described in INFCIRC/153 over 20 years ago, but fundamental changes to this system are now being explored. These changes would place even greater emphasis on searching for undeclared nuclear activities and thereby perhaps permit a significant reduction in the inspection of declared activities. Such measures as satellite surveillance and environmental monitoring techniques may hold promise, but that promise will need to be well demonstrated before the international community is willing to accept these measures instead of the existing system. These new measures will also stimulate new concerns of intrusiveness and compromise of sovereignty to be resolved. For the immediate future, IAEA safeguards will remain and are finally consistent with the vision of 1967-1972 and, to a lesser but reasonable degree, with the vision of 1946-1948.

Notes

1. William Epstein, *The Last Chance: Nuclear Proliferation and Arms Control* (New York: The Free Press, 1976), 4-6.
2. Ibid., 5.
3. Allan McKnight, *Atomic Safeguards: A Study in International Verification* (New York: UNITAR, 1971), 4.
4. Ibid., 5.
5. Most discussions tend to equate the Acheson-Lilienthal Report and the Baruch Plan. I am grateful to Leonard Spector for pointing out the significant differences between the two with respect to how they treated eventual U.S. entry into the envisioned new international regime and the consequence significant difference in acceptability to the USSR.
6. Ibid., 7.
7. Ibid., 8.
8. Ibid., 13.
9. Lawrence Scheinman, *The International Atomic Energy Agency and World Nuclear Order* (Washington, DC: Resources for the Future, 1987), 61-62. Disarmament efforts had been deadlocked from the beginning, and in the interim the USSR and the United Kingdom had detonated fission weapons, and the United States and the USSR had detonated fusion weapons. While the major powers had aggressive military programs, civil applications of nuclear energy were not being pursued with similar vigor.

10. The seven allies, selected on the basis that they were either advanced in nuclear technology or producers of nuclear materials, were the United Kingdom, Canada, France, South Africa, Belgium, Australia, and Portugal. See McKnight, 21-23, and Scheinman, 63-73.

11. Statute of the International Atomic Energy Agency ["Statute"], as amended up to 28 December 1989; Article II.

12. Statute, III/6.

13. That is, knowledge and "know-how" (whether in documentary forms such as blue-prints or not) as opposed to specific items of equipment. The IAEA would safeguard a specific piece of equipment, but did not address safeguarding replications of that equipment which the recipient country might fabricate.

14. Statute, III/B.4.

15. Ibid., V/D.

16. Ibid., V/E.3.

17. Ibid., XIX.B.

18. Ibid., VI/F.

19. Ibid., XII.C.

20. Historically, most North Korean nuclear cooperation has been with the Soviet Union, not the IAEA. The IAEA has provided training and assistance in such areas as surveying for uranium ore deposits and laboratory measurement techniques. Most of North Korea's nuclear facilities were built indigenously. Hence suspending cooperation had more political effect than power to impede the program.

21. Statute, VII/F.

22. Ibid., Article XII/A.1., 3., 5., and 6.

23. Ibid., Article XII/C.

24. Ibid., emphasis added.

25. Ibid., Article XI/F.

26. Originally all items subject to safeguards were either supplied through the Agency and hence captured under Article XI/F or made subject to safeguards by virtue of an agreement among the supplying state, the receiving State, and the Agency. The latter form of agreement was known as a Safeguards Transfer Agreement as it transferred the supplier's safeguards rights and responsibilities to the IAEA. When states began to make strictly voluntarily submissions, the practice was to negotiate an agreement, and the Secretariat appears to have just adopted the practice they were accustomed to.

27. McKnight, 46-53; quotation from 47.

28. IAEA document INFCIRC/66/Rev. 2, "The International Atomic Energy Agency's safeguards system, as provisionally extended in 1966 and 1968."

29. Scheinman, 131.

30. Ibid., 138. This decision had far reaching consequences for the right of non-nuclear weapon states to acquire "peaceful nuclear explosives." It effectively precluded any party to the Treaty of Tlatelolco—or other non-NPT safeguards obligations—from seeking to justify development of a nuclear explosive as being strictly for peaceful purposes and therefore permissible. In essence, the NPT approach to peaceful nuclear explosives, that the explosive be supplied by and used under the control of a nuclear weapon state, was extended to all safeguarded material.

31. Ibid., 139, 141; Ben Sanders, "IAEA Safeguards: A Short Historical Background," in *A New Nuclear Triad: The Non-Proliferation of Nuclear Weapons, International Verification and the International Atomic Energy Agency* (Southampton, England: Programme for Promoting Nuclear Non-Proliferation, 1992), 11.

32. Scheinman, 139.

33. McKnight, 67.

34. Ibid., 67-68.

35. Treaty on the Non-Proliferation of Nuclear Weapons ["NPT"], Article IX/3.

36. NPT, Article I.

37. NPT, Article II.

38. NPT, Article III/1.

39. NPT, Article III/1.

40. NPT, Article III.I. Actually, safeguards would apply to all nuclear material in all *peaceful* nuclear activities, as the negotiators of the NPT explicitly accepted the notion that some military activities, not associated with nuclear explosives, would be permitted. The convention did not discuss just what these "non-proscribed" military nuclear activities would be, although all recognized that using nuclear reactors to propel naval vessels was one. Interview with Myron Kratzer, former Atomic Energy Commission and State Department official, February 3, 1994.

41. This question was addressed in negotiating the NPT, and subseqently in formulating a new safeguards approach for implementing the NPT, but no formal understandings on this issue were ever documented. Essentially it was understood that using nuclear power to propell naval vessels was permitted, and no other nuclear uses of nuclear materials were foreseen (using depleted uranium in anti-tank ammunition was and is not considered a "nuclear" use). Interview with Myron Kratzer.

42. NPT, Article III/3.

43. NPT, Preamble. Article III/3 incorporates this language by reference into the provisions of Article III.

44. NPT, Article III/4.

45. Some states which are not NPT parties, such as India, argue that the IAEA Board of Governors is not responsible for or authorized to pursue a state which is flouting its NPT obligations by failing to conclude a safeguards agreement.

46. A more detailed description of this system can be found in "International Safeguards," by David Fischer, in *Safeguarding the Atom: A Critical Appraisal,* ed. Jozef Goldblat (London & Philadelphia: Taylor & Francis for Stockholm International Peace Research Institute, 1985), 79 ff.

47. The IAEA and member States have adopted the practice of using INFCIRC/153 as the model for other full-scope safeguards agreements as well, including those implementing the Latin-American nuclear-free zone (Treaty of Tlatelolco).

48. IAEA document INFCIRC/153, "The Structure and Content of Agreements Between the Agency and States Required in Connection with the Treaty on the Non-Proliferation of Nuclear Weapons," paragraph 14.

49. Interview with Myron Kratzer, former Atomic Energy Commission and State Department official, February 3, 1994.

50. INFCIRC/153, paragraph 28.

51. INFCIRC/153, paragraphs 29, 30.

52. Interview with Myron Kratzer. Kratzer has made this same point in several unpublished papers presented at meetings of the Atlantic Council's Project on Non-Proliferation.

53. INFCIRC/153, para. 73.

54. This statement appears in the Safeguards Statement which introduces the chapter on safeguards implementation in each IAEA annual report. See for example, *The Annual Report for 1992*, GC(XXXVII)/1060, printed by the International Atomic Energy Agency, July 1993, 135.

55. INFCIRC/153, para. 73.

56. See McKnight, for an excellent description of the efforts of many original members of the IAEA to maximize claims of sovereign immunity from intrusive inspections as soon as the new IAEA began operation in 1957, by refusing to fund the staff necessary to formulate a safeguards system and then by refusing to accept the Director General's efforts to respond to a Japanese request for safeguards on the grounds that his technical approach was not adequately studied and undergirded with technical justifications in detail.

57. Sanders, 9.

58. Unannounced inspections have been conducted only in situations in which the IAEA first negotiated specific arrangements with the member State to permit this practice, and these always involved special circumstances. One case involves inspections of cascade halls in centrifuge enrichment plants, and extensive negotiations between technology holders and the Agency over a complete package of inspection procedures, which ultimately included "limited-frequency unannounced-access" inspections." Another situation involved "short-notice random inspections" and arrangements to perform such inspections to test their utility and determine the problems which might arise.

59. The first manifestations of renewed political will actually occurred at the 1990 NPT Review Conference, where a number of states sought to re-emphasize the IAEA's right to perform special inspections.

60. See Lawrence Scheinman, "Lessons From Post-War Iraq For the International Full-Scope Safeguards Regime" and Maurizio Zifferero, "The IAEA: Neutralizing Iraq's Nuclear Weapons Potential," both in *Arms Control Today* 23, no. 3.

61. J. Christian Kessler, "History & Current Trends in Nuclear Safeguards," unpublished manuscript prepared for International Training Course on Implementation of State Systems of Accounting for and Control of Nuclear Material, May 12-28, 1993, Sante Fe, NM.

62. Sanders, 10.

63. Lawrence Scheinman, "The Current Status of IAEA Safeguards," in *A New Nuclear Triad*, 18.

64. Scheinman, "Lessons."

65. Article VI calls for "negotiations in good faith . . .on a treaty on general and complete disarmament under strict and effective interational control," but sets no timetable. Lack of progress over the past 25 years is one of the most contentious issues facing the parties today.

3. ❖ Biological Weapons

The story of international efforts to eliminate biological weapons is very different from that of nuclear weapons. The first efforts to prevent the use of biological weapons took place during the League of Nations' attempt to eliminate war, the horrors of which had been made so dismayingly manifest during the Great War. The 1925 Geneva Protocol addressed the use of both biological and chemical weapons, although only the latter had been used during World War I. Like most international agreements of that day, it did not include verification measures or otherwise address the issue of compliance.

Efforts to eliminate biological weapons took a further step in 1972 with agreement on the Biological Weapons Convention. Although the BWC was negotiated only shortly after the NPT, it is a very different treaty. Like the NPT, the BWC addresses possession rather than hostile use, but unlike the NPT the BWC does not include any measures explicitly dealing with judging compliance (although it does deal with resolving concerns of one state party regarding the compliance of another). This may stem from the very different international legal and political pedigrees of the two conventions: the NPT is the culmination of efforts started at the founding of the United Nations with the UNAEC and the Baruch Plan, while the BWC continues the heritage of the Geneva Protocol. Alternatively, the differences may be due to more substantive and technical factors rather than political ones, arising from the relative difficulties in verifying compliance in each case. This is a significant issue, and building a stronger system to demonstrate and determine compliance has been the central theme of BWC review conferences since the convention entered into force.

The Geneva Protocol

The first international effort to eliminate biological weapons[1] was made in 1925, in the context of a larger disarmament effort. The Geneva Protocol ("Protocol for the Prohibition of the Use in War of

Asphyxiating, Poisonous or Other Gases, and of Bacteriological Methods of Warfare") was written at the 1925 Geneva Conference for the Supervision of the International Traffic in Arms. The Geneva Protocol contains no provisions addressing enforcement, it being assumed that the deterrent implications of no first use would prove adequate. Until a 1988 Security Council Resolution established an explicit legal relationship for purposes of enforcement, the Geneva Protocol was widely viewed as more hortatory than effective.

The Protocol specifically prohibits "the use of bacteriological methods of warfare" and states parties are "bound as between themselves according to the terms of this declaration."[2] While the ban on use of "bacteriological" (the term used at that time) weapons against other parties is clear, the references to chemical weapons are less so (this issue will be addressed later). The Protocol itself contains no verification or other compliance or enforcement related provisions; the concept of including compliance related provisions into an international convention would become common only following World War II. More importantly, at that time it was presumed that any first use (or at least militarily significant first use) would be obvious, and the perpetrator clear.

This is not the place to explore the strong and weak points of the Protocol with respect to the obligations it creates for parties; what is important for this study is that it firmly establishes several ideas. First, there is the normative point that biological and chemical weapons are not acceptable weapons among modern civilized nations. Second, it reiterates the principle that international law may prohibit the use of an entire class of weapons (treaties concluded before World War I had established prohibitions against use of chemical weapons, but these treaties were repeatedly violated by all major belligerents during the war). Finally, it establishes a clear link between two kinds of "weapons of mass destruction," a link that plays an important role in later developments. At the same time, the Geneva Protocol left vital work undone; while use was prohibited (actually, only first use against another state party), possession was not. But the potential for a nonparty, or a party acting in violation of its obligation, to attack with biological (or chemical) weapons meant that each party would be remiss in protecting itself unless it possessed these weapons. It appears that for compliance verification to become an important and routine part of a treaty, possession must either be prohibited or limited in some measurable fashion. When only use is prohibited, compliance questions arise only when illegal use is alleged, and historically the matter has required political rather than technical resolution.

Negotiating the Biological Weapons Convention

The Convention was negotiated in the Conference of the Committee on Disarmament (CCD) between 1969 and 1972. Compared with the many years it took to negotiate the NPT, or would take to negotiate the CWC, the BWC was negotiated very quickly. There are several views on why this was possible.

During this period the United States and the Soviet Union had, in the judgment of at least one observer, made progress on strategic arms control a "superordinate" goal, and the desire to maintain momentum on the larger arms control agenda may be the reason for "the sudden pressure for an early agreement—even on rather less than satisfactory terms—that accelerated the CCD's proceedings in mid-1971 to the detriment of the Convention in general and of its more robust British ingredients in particular."[3] (The British draft text contained significant compliance and verification related provisions.) Nicholas Sims, perhaps the principal student of the BWC's development, hypothesizes that Moscow and Washington viewed the BWC as a means to maintain momentum on arms control, "to find yet one more area in which the USA and the USSR shared a common interest in restraint."[4]

The BWC does not contain any verification mechanism, much less a complex and intrusive international inspection regime such as IAEA safeguards. Sims also argues that, in addition to the "sudden" pressure to conclude the convention, the absence of verification provisions is due to a belief that biological weapons were not really credible military weapons because of the difficulties in controlling the effects of such weapons once employed, plus the complicated requirements for protecting one's own forces. In 1969 President Nixon unilaterally terminated U.S. offensive biological weapons programs, arguing that biological weapons are not militarily useful. Many Western and nonaligned states were willing to accept the absence of verification measures only for this reason, and to focus attention on the more militarily pressing matter of controlling chemical weapons. It is clearly the case that, because the United States had publicly foresworn biological weapons, the administration was inclined to seize the opportunity to obtain legally binding pledges from other states, with or without verification.[5] Moscow may have had a second reason for wishing to avoid the verification issue, an intention to violate the treaty.[6]

Another factor for many participants was difficulty in identifying any multilateral verification measures that would be credible (implying that the British proposals did not). At this time the U.S. technical community's thinking was still focused on national technical means.[7] Finally, both chemical and biological weapons were "on the agenda" of the CCD, and ACDA's history of the BWC notes that while the United States did not consider either to be of greater priority than the other, "it held that biological weapons presented less intractable problems" and an agreement on biological weapons should not be delayed until "reliable prohibition" of chemical weapons could be agreed upon.[8]

The Convention on the Prohibition of the Development, Production and Stockpiling of Bacteriological (Biological) and Toxin Weapons and on Their Destruction (BWC) was signed in April 1972 and entered into force in March 1975 following ratification by the three depositary states (the United States, the Soviet Union, and the United Kingdom). Article I prohibits development, possession, or retention of biological agents and toxins "of types and in quantities" without justification for peaceful purposes, as well as the weapons or other means for using such agents or toxins "for hostile purposes or in armed conflict." Article II obligates parties to destroy all such weapons they may possess. Article III prohibits transfer to "any recipient whatsoever" and prohibits assistance to other states (not limited to parties) or international organizations regarding manufacture or acquisition of agents, toxins, or means of delivery.

The Convention does not specifically address verification, but it does address the question of noncompliance. First, parties "undertake to consult with one another and to cooperate in solving any problems which may arise in relation to the objective of, or in the application of the provisions of, the Convention." Parties may also avail themselves of "international procedures within the framework of the United Nations and in accordance with its Charter."[9] Article VI provides that any party "which finds that any other State Party is acting in breach of obligations" may "lodge a complaint with the Security Council." Such a complaint "should include all possible evidence confirming its validity." In addition, each party "undertakes to cooperate in carrying out any investigation which the Security Council may initiate, in accordance with the provisions of the Charter of the United Nations, on the basis of the complaint received by the Council." Furthermore, "The Security Council shall inform the States Parties to the Convention of the results of the investigation." Finally, each party "undertakes to provide or support assistance, in accordance with the United Nations Charter, to any party to the Convention which so requests, if the Security Council decides that such Party has been exposed to danger as a result of violation of the Convention."[10]

Ratification Through the First Review Conference

During the early years no party formally accused another of violating BWC obligations, although inklings of informal accusations against the Soviet Union came from the United States. However, as Sims notes, there were no formal consultations between the two superpowers regarding these concerns, as is provided for by Article V. Events during the first review conference would build on this lack of regular procedure.

The First Review Conference was held in March 1980 under rather inauspicious circumstances. Ten weeks earlier the USSR had invaded Afghanistan, and international concern over the outbreak of anthrax in Sverdlovsk the year before was building rapidly.[11] During this first review conference, reference to "strengthening" the Convention was "virtually taboo."[12] Nonetheless, this First Review Conference did make some progress in clarifying the terms of the Convention and in specifying a consultative procedure.[13]

The Declaration of the First Review Conference notes the confidence-building value of voluntary declarations by parties concerning their past biological weapons programs and steps to eliminate such programs.[14] The Declaration also "invites" parties to submit texts and information concerning national legislation or regulations implementing Convention obligations to the U.N. Secretariat.[15] More importantly, in its Final Declaration, the Conference observed that this "enables interested States Parties to use various international procedures which would make it possible to ensure effectively and adequately the implementation of the Convention provisions."[16] This clarification emphasized that a party dissatisfied with the results of bilateral efforts to resolve a compliance concern "could bring it before the collectivity of States Parties, represented by their experts in a veto-free setting."[17] This has the advantage that a party "would no longer face the stark choice between bringing a formal complaint to the Security Council (under Article Six) where even its consideration, let alone investigation and findings, might well be prevented, and making accusations subject to no organized international scrutiny."[18] Compliance concerns could now be pursued without jeopardizing the treaty regime itself (to the degree that accusations of violations that cannot be or are not resolved tend to undermine confidence in the treaty).[19]

This apparently bland language concerning consultation regarding allegations of violations concealed a struggle pervading the Conference,and threatening to derail it. Sweden entered the Conference intent on pursuing amendment of the BWC "to separate the factfinding stage from the adjudication on a complaint" so that a permanent member of the Security Council could not prevent investigation of allegations against it.[20] To do this, Sweden proposed establishing a Consultative Committee of Experts, modeled on the ENMOD Convention.[21] As the Review Conference struggled with the appropriateness of amending the Convention, and how otherwise to address this issue, newspapers reported that the United States had formally requested an explanation of the Sverdlovsk anthrax outbreak from the Soviet Government. These news accounts appeared to catch the U.S. delegation by surprise and certainly changed the tenor of Conference proceedings.[22]

The Soviet delegation rejected any amendment of the Convention, as did many other delegations, and also rejected the need for any improvement in or explanation of the consultative process referred to in Article V. British ambassador Summerhayes responded by pointing out that not only did the ENMOD Convention provide for a consultative committee of experts, but the two superpowers had the previous year jointly submitted a draft convention on radiological weapons which included an experts committee to investigate alleged violations.

In the end, a compromise was worked out among the British, Swedish, and Soviet ambassadors. This compromise consisted of a commentary in the Final Declaration elaborating on the right of any state party "to request that a consultative meeting open to all States Parties be convened at expert level."[23] What the Final Declaration did not identify was the entity to which such a request was to be made. This reflected strong disagreement between the Swedish faction, which insisted upon the U.N. Secretary-General, and the Soviet bloc, which would only accept the depositary governments as conveners of such a meeting. In his concluding statement on this text, Ambassador Summerhayes explained that the expert meeting could be convened "by Depositaries," so as to imply that it was not necessary for all three depositaries to agree on the meeting.[24] A permanent member of the Security Council now could not prevent an investigation, although it could still impede it (for example, by refusing to participate or cooperate, notwithstanding the Article V commitment to cooperate).

Finally, reflecting the strong historical link between biological and chemical disarmament efforts, the Conference expressed deep regret that agreement has not been reached on "effective measures for the prohibition of the development, production and stockpiling of chemical weapons and their destruction" pursuant to Article IX, and urged further negotiation on this matter.

Before the Second Review Conference

In 1982 several efforts were made to strengthen compliance measures. At the U.N. Second Special Session on Disarmament Sweden introduced a resolution containing the language negotiated for the Review Conference's Final Declaration and calling for a special conference to review verification of the BWC, obviously with a view to amendment. This resolution passed, and subsequently the United States and the United Kingdom circulated a Swedish note requesting the conference. However, they received little formal response, reflecting concerns over the value of a conference that the USSR and its allies would refuse to attend, even if the USSR (acting in its role of third depositary) did not seek to prevent it from taking place. Just as at the Review Conference, many supporters of strengthening the BWC compliance procedures were also very concerned that they not create a two-tiered treaty, with some parties committed to procedures explicitly rejected by others.

Belgium took a very different approach, introducing at the Special Session an entirely procedural draft treaty. Linked by its title and preamble to both the BWC and the Geneva Protocol, the Belgian proposal would commit all parties to accept U.N. factfinding with respect to their compliance to the two substantive treaties. While an innovative approach, this one too would create two categories of states with respect to compliance, and for this reason it, too, failed.[25]

Acting at the next General Assembly session, France took yet another approach. France's resolution requested the Secretary-General

> to investigate, with the assistance of qualified experts, information that may be brought to his attention by any member State concerning activities that may constitute a violation of the Protocol or of the relevant rules of customary international law in order to ascertain thereby the facts of the matter, and promptly to report the results of any such investigation to all Member States and to the General Assembly.[26]

The French resolution did not refer to the BWC, as France was not a party. This created a potential defect, depending on whether the reference to customary international law was interpreted to include possession of biological weapons. Also noteworthy is that investigative results are to be reported to the General Assembly rather than the Security Council. While the Security Council would retain primary jurisdiction (which it has for all threats to international peace and security), any veto by a permanent member would be conspicuous. The French resolution passed the General Assembly in December 1982, but has yet to be implemented. As Nicholas Sims notes, the "very existence of new procedures for the investigation of BW use allegations" could, if implemented, "alter significantly the pattern of expectation" when BWC parties again consider strengthening Article V.[27]

For whatever reason, the United States did not press the issue of the Sverdlovsk anthrax outbreak until the Reagan administration, shortly after taking office in January 1981, began making allegations of Soviet development and offensive use of biological warfare agents. Allegations that the Sverdlovsk anthrax outbreak was caused by an explosion at an illegal biological weapons facility were soon followed by claims that the USSR was using biological (mycotoxin) agents in Laos, Kampuchea, and Afghanistan. Lastly, the Reagan administration, based on information provided by emigrés, accused the USSR of pursuing genetic engineering technologies to develop new biological weapons.[28]

Second Review Conference

These U.S. accusations of Soviet violation of the Convention (and, in the "yellow rain" case, of the Geneva Protocol) pervaded the Second Review Conference, creating a "mood of uncertainty."[29] However, the accusations were widely perceived to be very shaky, as efforts to replicate the original findings proved difficult.[30] Even Canada, the strongest supporter of the U.S. allegations, noted that its own investigation could not resolve the matter and expressed regret that a thorough examination had not been possible.[31] The Irish delegate stated "that unless means were found to deal objectively with such allegations, erosion of the authority of the Convention might well be inevitable."[32]

Several participants, including the Soviet Union, proposed negotiating a separate protocol containing verification measures. While the USSR's interest in negotiating a separate protocol (the possible contents of which the USSR did not outline) may have reflected a desire to defer the verification issue, it is clear that many states were concerned at the lack of institutional mechanisms for addressing and resolving the U.S. accusations.[33] The United States, among others, argued for postponing negotiations on a verification protocol, citing the ongoing negotiations on a chemical weapons convention and expressing the concern that concurrent negotiations would distract the negotiators and delay progress on the CWC.[34]

Concerns about the technical feasibility of adequate verification also motivated the United States.[35]

The Final Declaration of the Second Review Conference reflects this controversy. In an effort to strengthen the Convention, the Conference made several significant interpretations of the treaty, including a statement that "the Convention unequivocally applies to all natural or artificially created microbial or other biological agents or toxins whatever their origin or method of production." The undertakings of the Convention were specifically interpreted to include not only biologically produced agents but "their synthetically produced analogues."[36] While political links with a convention prohibiting chemical weapons had long been made, this interpretation explicitly established an overlap in the scope of the two conventions, extending coverage of the BWC until a CWC could be concluded.

The first direct discussion of verification issues comes under Article V (which addresses consultations and cooperation among states parties to solve problems arising with respect to the objective or implementation of the BWC). In addition to noting again (as did the Declaration of the First Review Conference) that consultations and cooperation may be undertaken in accordance with the U.N. Charter, consultative meetings may be convened at the expert level. The Declaration "stresses the need for all States to deal seriously with compliance issues and emphasizes that the failure to do so undermines the Convention and the arms control process in general."[37] The Declaration expands the scope of "consultative meetings" to include

> any problems which may arise in the relation to the objective of, or in the application of the provisions of the Convention, suggest ways and means for further clarifying, *inter alia*, with assistance of technical experts, any matter considered ambiguous or unresolved, as well as initiate appropriate international procedures within the framework of the United Nations and in accordance with its Charter.[38]

More importantly,

> The Conference considers that States Parties *shall* [emphasis added] co-operate with the consultative meeting in its consideration

of any problems . . . and in clarifying ambiguous and unresolved matters.[39]

In addition to considering how to investigate compliance when one party casts doubt on the compliance of another, the Conference also addressed how to increase transparency. The Conference acted "to strengthen the authority of the Convention and to enhance confidence in the implementation of its provisions," by stipulating that parties "*are* to implement" (emphasis added) specified measures "in order to prevent or reduce the occurrence of ambiguities, doubts and suspicions."[40] Four broad measures were identified:

- Exchange of data on research centers and laboratories that meet appropriate safety standards for handling biological materials relevant to biological weapons
- Exchange of information on all outbreaks of infectious diseases and similar occurrences caused by toxins, that seem to deviate from the normal pattern
- Encourage publication of the results if biological research directly related to the Convention in scientific journals widely available to the international community
- Actively promote contacts among scientists engaged in biological research directly related to the Convention.

An ad hoc meeting of "scientific and technical experts" from the parties was directed to finalize the modalities for data exchange.

While the BWC permits any party to bring concerns about noncompliance to the Security Council, the Second Review Conference expanded the scope of outside appeals and assistance. Referring again to the "need to further improve and strengthen this and other procedures to enhance greater confidence in the Convention," the Conference expressed the view that the Security Council "may, if it deems it necessary, request the advice of the World Health Organization in carrying out any investigation of complaints lodged with the Council." As the BWC does not establish any separate and permanent Secretariat or technical staff, this is the first effort to identify a permanent international staff expert in biological and disease matters that may act to investigate concerns regarding non-compliance with the Convention. While only a rather small step towards compliance verification, this decision, coupled with the information exchange and transparency measures discussed under Article V, represents serious efforts to address the verification problem.

The Second Review Conference has been judged "a surprising success," especially so in light of the several U.S. allegations of Soviet violations of the Convention and the Soviet responses, which were not particularly forthcoming.[41] Sims states that this Review Conference "did much to arrest the seven-year decline in credibility of the comprehensive ban on biological weapons. . . . That treaty regime had never been robust, and it had suffered an alarming erosion of confidence, particularly in the United States."[42] He attributes this to several factors: "Its verification procedures are minimal; its consultative mechanisms are underdeveloped; it is not equipped with regular scientific advice, let alone with any permanent institutions or secretariat."[43] Which is to say, there were no effective mechanisms for addressing U.S. allegations that the Soviet Union was violating the treaty.

The decisions taken at the Second Review Conference reflected a commitment on the part of the participating governments to develop remedies for some of these weaknesses. In his concluding summary the President of the Conference observed that this review conference took steps "beyond the traditional function of scrutinizing the past performance of a treaty . . . it had to innovate . . . it had to strengthen an ailing treaty regime without the possibility of major surgery."[44]

Third Review Conference

The Second Review Conference had accomplished several important steps. In addition to establishing a procedure for investigation and evaluating accusations of noncompliance in a less politicized and confrontational forum than the Security Council, it had also imposed various reporting requirements on states parties, the modalities of which were determined by the 1987 ad hoc group of scientific and technical experts. It is difficult to evaluate the degree to which States began to report unusual disease outbreaks to WHO, but if performance on reporting biological research and production facilities to the U.N. Disarmament Affairs Department is indicative, the record is not good. While states were to submit initial reports for 1986, and subsequently each year, by the end of 1988 only 21 percent had reported. Certainly most states parties had no facilities to report, but even a "null" return was appropriate and required; only New Zealand, it seems, filed one.[45] When the Third Review Conference convened, much remained to be done, even in terms of implementing the measures agreed upon 5 years earlier.

Meeting in September 1991, the Third Review Conference picked up where the Second had left off. The discussion of Article I in the Final Declaration returns to the central theme of the previous two conferences, as it notes that the Conference:

> expresses concern at statements by some States parties that compliance with Articles I, II and III has been, in their view, subject to grave doubt in certain cases and that efforts since the Second Review Conference to resolve these problems have not been successful.[46]

For the most part, the Final Declaration of this conference reiterates or reaffirms the proposals, interpretations, and recommendations included in the Final Declaration of the previous conference. The new ground is represented by a set of confidence-building measures identified under Article V. These measures build on those agreed at the previous review conference and at the 1987 Ad Hoc Meeting of Scientific and Technical Experts, but also extend the compliance related elements of the regime. The Conference, according to the Final Declaration, "agrees that the States parties are to implement, on the basis of mutual cooperation, the following measures . . . to prevent or reduce the occurrence of ambiguities, doubts and suspicions. and in order to improve international cooperation in the field of peaceful bacteriological (biological) activities."[47] These measures are:

- Exchange of data on research centers and laboratories
- Exchange of information on national biological defense research and development programs
- Exchange of information on outbreaks of infectious diseases and similar occurrences caused by toxins
- Encouragement of publication of results and promotion of use of knowledge

- Active promotion of contacts
- Declaration of legislation, regulations and other measures
- Declaration of past activities in offensive and/or defensive biological research and development programs
- Declaration of vaccine production facilities
- Annual declaration of "nothing to declare" or "nothing new to declare" when appropriate.[48]

A second important development is clarification of the entity responsible for convening a "working group at expert level" to investigate and report on allegations of violations. The Final Declaration notes that a formal consultative meeting "could be preceded by bilateral or other consultations by agreements among those States parties involved in the problem," clearly implying that states are to be prudent in use of the formal mechanism and indicating the view that informal consultations between accuser and accused are less damaging to the regime than immediate recourse to the formal process, whether or not the accusations prove valid. The Conference then addresses the question left hanging at the previous review conference— who is to convene a formal meeting to conduct this factfinding:

> Requests for the convening of a consultative meeting shall be addressed to *the* Depositaries, who *shall immediately inform* all States parties of the request and *shall convene* within 30 days an informal meeting of interested States parties to discuss the arrangements for the formal consultative meeting, *which shall be convened* within 60 days of receipt of the request."[49]

The Soviet Union won on the question of who convenes (all Depositaries), but discretion is clearly eliminated ("shall immediately inform . . . ; shall convene"), so that a Permanent Member of the Security Council (as all the Depositaries are) cannot block the factfinding phase. The Swedes (and the Western group) won the larger point. Undoubtedly this reflected the new attitude in Moscow that led to so many fundamental changes and ultimately the dissolution of the Soviet Union. More importantly, for the first time there is a clearly established and credible mechanism for addressing concerns about the compliance of a party.

What remained unresolved was whether there could be any politically neutral, international mechanism for performing routine verification of compliance by all parties. This issue, too, was addressed by the Conference. In its Final Declaration the Conference recorded its decision "to establish an Ad Hoc Group of Governmental Experts open to all States parties to identify and examine potential verification measures from a scientific and technical standpoint."[50] That is, the group was to act as experts evaluating the technical capabilities of possible verification measures, and not as government representatives negotiating on what verification should be implemented. The Ad Hoc Group's terms of reference were spelled out in some detail in the Final Declaration itself. They included identifying measures that "could determine whether a State party is developing, producing, stockpiling, acquiring, or retaining" either (a) biological agents or toxins "of types and in quantities" without credible peaceful uses, or (b) "weapons, equipment or means of delivery designed to use such agents or toxins for hostile purposes or in armed conflicts."[51] The Group was directed to examine measures that might prove useful either singly or in combination, and to evaluate these measures in terms of six criteria:

- Their strengths and weakness based on, but not limited to, the amount and quality of information they provide, and fail to provide
- Their ability to differentiate between prohibited and permitted activities
- Their ability to resolve ambiguities about compliance
- Their technology, material, manpower, and equipment requirements
- Their financial, legal, safety, and organizational implications
- Their impact on scientific research and cooperation, and industrial development, and their implications for the confidentiality of commercial proprietary information.[52]

This Ad Hoc Group was to perform a rigorous and complete study of verification measures and by implication, of verification requirements, which is without equal in the nuclear arena, and probably, at this point, in the chemical arena. The work of this Ad Hoc Group will be described later.

The Third Review Conference also returned to several themes touched on by the previous review conferences. The obligation of all parties "to provide a specific, timely response to any compliance concern alleging a breach of their obligations under the Convention" had been discussed previously, but the new language stipulates detail and timeliness not in the Second Review Conference's Final Declaration. While the previous conference had alluded to the competence of the U.N. Secretary-General, this time the Conference took note of Security Council Resolution 620 of 1988, "which encouraged the United Nations Secretary-General to carry out prompt investigations, in response to allegations brought to his attention by any Member State concerning the possible use of chemical and bacteriological (biological) or toxin weapons."[53] The Conference even referred obliquely to the Security Council's extensive powers under Chapter VII: "The Conference stresses that in the case of alleged use the United Nations is called upon to take appropriate measures, which could include a request to the Security Council to consider action in accordance with the Charter."[54]

If all these steps were fully implemented, the BWC would become a disarmament treaty with significant teeth. This result stands in stark contract to the character of the BWC at its birth and would reflect the first time that a nonproliferation convention was amended in a major fashion. The question remains as to how any recommendations from the Ad Hoc Group might be implemented so that a two-tiered treaty does not result. While no obvious answer is available, a great deal of creativity has already been demonstrated on behalf of the BWC, and perhaps a future review conference will identify a way.

VEREX

The Ad Hoc Group of Governmental Experts to Identify and Examine Potential Verification Measures from a Scientific and Technical Standpoint, which became known as "VEREX," met four times during 1992 and 1993. During nearly 8 weeks of work, the group identified 21 individual measures; categories of offsite measures include information monitoring methods, data exchange measures, remote sensing technologies, and inspection related activities; categories of onsite measures included exchange visits, inspection techniques, and continuous monitoring technologies. Each measure was evaluated against the criteria identified by the Third Review Conference in terms of its relevance for judging compliance with prohibitions regarding development, acquisition, and stockpiling of biological weapons. Representative combinations of these measures were also considered to evaluate the potential for synergy's among measures:

> [VEREX] concluded that potential verification measures as identified and evaluated could be useful to varying degrees in enhancing confidence, through increased transparency, that States Parties were fulfilling their obligations under the BWC. While it was agreed that reliance could not be placed on any single measure to differentiate conclusively between prohibited and permitted activity and to resolve ambiguities about compliance, it was also agreed that the measures could provide information of varying utility in strengthening the BWC. It was recognized that there remain a number of further technical questions to be addressed such as identity of agent, types and quantities, in the context of any further work. Some measures in combination could provide enhanced capabilities by increasing, for example, the focus and improving the quality of information, thereby improving the possibility of differentiating between prohibited and permitted activities and of resolving ambiguities about compliance.[55]

Readers with experience in multilateral fora will recognize the classical result of posing the question whether the glass is half full or half empty—"There is water in the glass." In this situation (as in many others) there is some justification for such a result.

VEREX was charged with judging the potential utility of individual measures which might be utilized were efforts made to create a verification regime, but the responsibility for negotiating such a regime and the question of whether effective verification is even possible were beyond the mandate of this group. While many states seem confident that adequate verification of the BWC is possible, a few, led by the United States, hold reservations about the capabilities of available technical measures to provide credible verification. In response to these concerns, the VEREX final report carefully avoids any reference to verification, and speaks in terms of "enhancing confidence."

VEREX must be judged a success, as it provided an extensive evaluation of measures which are the building blocks for any verification system. The conclusions drawn by VEREX were, per its mandate, arrived at by consensus. The question remaining is whether some set of these measures (with the possible but unlikely addition of as yet unidentified measures) can be assembled into a verification regime satisfying the technical and political requirements of a majority of the parties. Perhaps the situation was best summarized in a working paper prepared by the United Kingdom and submitted the first day VEREX met:

> As the primary operational objective is to detect non-compliance, candidate regimes should be evaluated in terms of their ability to achieve a reasonable probability of detecting non-compliance in a reasonable proportion of evasion scenarios. However, a verification regime aimed at detecting evasion that even performs only moderately well is likely to have a significant deterrent effect on potential evaders. The wider the range of the national activities, whether research, industry, public and veterinary health, or military, that are subject to scrutiny by a BWC regime, the harder it is likely to be for an evader to conceal a BW programme. the higher the risk of discovery, and the greater the deterrence.[56]

Analysis

The BWC originated in a very different political context than did the NPT. First, unlike the NPT, the BWC is not a two-tiered treaty, permitting some powers (which in the NPT case happened to coincide in the main with the permanent members of the Security Council) to possess biological weapons while prohibiting them for all others. Any verification measures would have to be acceptable to the Permanent Five not only in terms of the confidence produced, but also in terms of the intrusions of national sovereignty required. In 1972 the Soviet Union had repeatedly and explicitly rejected any kind of onsite inspections for treaty verification. Because Soviet objections were a foregone conclusion, the United States had never been forced to address the question of accepting obligatory third party onsite inspection of an entire industry.[57]

Second, the United States had determined that biological weapons were in fact of no real military value and had publicly divested itself of such weapons on this basis. This judgment was not universally shared and in any case applied only to battlefield use and not to all possible applications. Nonetheless, it influenced how issues of verification were approached in important respects. In light of President Nixon's statement, it could be argued that verification of compliance was less pressing as states would not violate their obligations and surprise an adversary with so unpredictable a weapon.[58] Verifying noncompliance (that is, manufacture and possession) of the very small quantities of agent needed for some other applications (e.g., covert use against opposing leadership) were widely judged to be technically impossible. More generally, there were (and remain) many difficult technical questions about whether credible verification of nonpossession undertakings is possible.

Finally, there appears to have been a recognition within the U.S. Government that no multilateral organization would be able to provide effect verification of Soviet compliance and take meaningful enforcement actions. The issue would, by the character of the international situation, depend on U.S. monitoring of Soviet compliance and effective marshalling of international opinion in the event violations were detected. The superpowers were beyond the capabilities of international institutions to control.

For all these reasons the United States and the Soviet Union judged that establishing an international organization charged with verification and enforcing compliance with the BWC was undesirable, or at least unnecessary. Sweden and a few Western allies (including Britain) did not agree with this assessment and sought to redress the situation at the first review conference. While they essentially failed, by the Second Review Conference a growing number of parties were concerned about the absence of any international mechanism for addressing compliance concerns, driven in large measure by concern that the U.S. accusations against the Soviet Union were seriously undermining confidence in the BWC and hence eroding the established norms. By the Third Review Conference these concerns were buttressed by new ones. The Gulf War led to the discovery that

Iraq had an active offensive biological weapons program, countering any remaining sense that these weapons were militarily worthless for all countries. Whatever the validity of the "poor man's nuclear weapon" argument, it was clear that some countries believed and were acting upon this notion. In addition, biotechnology and engineering had progressed to the point where many countries were concerned with the possibility that an adversary could either modify a known pathogen so that existing vaccines would not be effective, or create a new form of pathogen.[59]

Whether VEREX will result in the addition of verification measures to the BWC remains to be seen. A special conference to consider the VEREX report is scheduled for September 1994. There appears to be a strong interest among governments to act on the VEREX conclusions and to assign verification responsibility to some international organization. Among the possible organizations that might be charged to perform BWC verification are the World Health Organization (converting it to something like the IAEA model); the Organization for the Prohibition of Chemical Weapons (now being established in The Hague pursuant to the CWC); or some element of the United Nations itself. The latter might involve the Department on Disarmament Affairs, the U.N. Special Commission (a temporary organization created by Security Council Resolution 687 to disarm Iraq), the Secretary-General drawing inspectors from lists of qualified individuals proposed by governments (as he is empowered to do for a wide range of investigations), or the creation of a new entity within the United Nations.[60] Each has its early proponents, but until the international community determines what verification measures are to be pursued, the entity to perform the verification remains speculation.

Perhaps the most intriguing point is how the BWC compliance system will develop further. As noted, the BWC originated in a very different context from the NPT but sharing common roots with and strong political links to a chemical weapons convention. It now appears possible that the BWC compliance system might develop into a regime quite similar to the system now being established for the recently signed Chemical Weapons Convention (to be discussed in the next chapter), and maybe even similar to the NPT compliance system (assuming continued evolution on the part of the later).

Notes

1. The term "biological" is used broadly to include all uses of microorganisms, including potentially prions (quasi-organic entities smaller than viruses), as well as organically derivable compounds—toxins—whether organically or synthetically manufactured. There is a substantial literature explicating the question of what are and are not "biological" weapons. No effort to unpack this issue will be made here.

2. Protocol for the Prohibition of the Use in War of Asphyxiating, Poisonous or Other Gases, and of Bacteriological Methods of Warfare ["Geneva Protocol"], signed in Geneva, Switzerland, June 17, 1925.

3. Nicholas A. Sims, *The Diplomacy of Biological Disarmament: Vicissitudes of a Treaty in Force, 1975-85* (New York: St. Martin's Press, 1988), 34.

4. Ibid., 36.

5. Interview with Charles Van Doren, former career official in and ultimately assistant director for nuclear and nonproliferation policy in the Arms Control and Disarmament Agency, July 25, 1994.

6. See, for example, Arkady N. Shevchenko, *Breaking with Moscow* (New York: Alfred A. Knopf, 1985), 173-174. "The Politburo approved this approach. The toothless convention regarding biological weapons was signed in 1972, but there were no international controls over the Soviet program, which continues apace," 174.

7. Interview with senior official, U.S. Arms Control and Disarmament Agency, 29 October 1993.

8. U.S. Arms Control and Disarmament Agency, *Arms Control and Disarmament Agreements* (Washington, DC: GPO, 1990), 130.

9. Convention on the Prohibition of the Development, Production and Stockpiling of Bacteriological (Biological) and Toxin Weapons and on Their Destruction ["BWC"], Article V.

10. BWC, Article VII.

11. Nicholas A. Sims, "The Second Review Conference on the Biological Weapons Convention," in *Preventing a Biological Arms Race,* ed. Susan Wright (Cambridge: The MIT Press, 1990), 268.

12. Ibid., 268.

13. Ibid., 268.

14. *Final Declaration of the First Review Conference*, Article II. (repr., Susan Wright).

15. Ibid., Article IV.

16. Ibid., Article V.

17. Sims, "Second Review," 68.

18. Ibid., 269.
19. Ibid.
20. Sims, *Diplomacy*, 100-101.
21. Convention on the Prohibition of Military or any other Hostile Use of Environmental Modification Technologies (May 1977).
22. Ibid., chap. 7, 8.
23. *Final Declaration of the First Review Conference*, Article V.
24. Sims, *Diplomacy*, 168-190.
25. Ibid., 212-213.
26. Ibid., 214.
27. Ibid., 216.
28. Leonard Cole, "Sverdlovsk, Yellow Rain, and Novel Soviet Bioweapons: Allegations and Responses," in Susan Wright, 199, 209-211. See also Milton Leitenberg, "Anthrax in Sverdlovsk: New Pieces to the Puzzle," in *Arms Control Today* 22, no. 3, 10-13.
29. Cole, 215.
30. Ibid., passim.
31. Ibid., 216.
32. Quoted in Cole, 216.
33. Sims in Wright, 269-271, and Barbara Hatch Rosenberg and Gordon Burck, "Verification of Compliance with the Biological Weapons Convention," in Wright.
34. Ibid., 301.
35. Interview with senior ACDA official, October 29, 1993.
36. *Final Declaration of Second Review Conference*, Article I (repr., Wright).
37. Ibid., Article V.
38. Ibid.
39. Ibid.
40. Ibid.
41. Sims in Wright , 267.
42. Ibid., 267.
43. Ibid.
44. Quoted in Rosenberg & Burck, 302.
45. Sims in Wright, 282-283; interview with senior ACDA official, October 29, 1993.
46. *Final Declaration of the Third Review Conference*, BWC/CONF.III/23, Geneva, 1992, Article I.
47. Ibid., Article V.
48. Ibid., Annex.
49. Ibid., Article V, 15, emphasis added.

50. Ibid., 16.
51. Ibid., 17.
52. Ibid., 17.
53. Ibid., 18.
54. Ibid., 18.
55. VEREX/8, paragraph 31.
56. VEREX/WP.1, paragraph 5.
57. The United States permitted IAEA inspection of a few government-owned nuclear facilities during the 1960s, but these were demonstration and test inspections, not routine implementation of an industrywide obligation to accept verification.
58. However unpredictable a battlefield weapon biological weapons may be, this has not kept a number of states from establishing covert programs. Estimates vary but most unofficial observers and U.S. Government estimates tend to run in the low to mid-twenties.
59. Interview with senior ACDA official, October 29, 1993.
60. Ibid.

4. ❖ Chemical Weapons

Efforts to prohibit chemical weapons date back to the end of the 19th century. The International Peace Conference held in The Hague in 1899 included in its convention a prohibition against asphyxiating gases along with dum-dum bullets and the launching of weapons from balloons. The prohibition against asphyxiating gases was repeated in the 1907 Hague convention.[1] During World War I these prohibitions were ignored as all major belligerents used chemical weapons on the battlefield. The end of the war saw a renewal of these arms control efforts, and during the 1925 Geneva Conference for the Supervision of the International Traffic in Arms, the United States proposed that the export of gases for uses in war be prohibited.[2] France countered with a proposal that use of any poisonous gas in war be prohibited, and it was this French proposal that became the Geneva Protocol. Reflecting the prewar conventions of 1899 and 1907, the Geneva Protocol begins with the observation, in its preamble, that "the use of asphyxiating, poisonous or other gases, and of all analogous liquids, materials or devices, has been justly condemned by the general opinion of the civilized world." Some observers have taken this language to indicate a judgment that the use of chemical weapons in war violates customary international law, as opposed to treaty law.[3] It certainly goes beyond the terms of the 1899 agreement (which prohibits only first use) and the 1907 Hague agreement, both of which are limited to actions between parties.

The Geneva Protocol's prohibition against first use stood, categorical but without any defined, much less effective, enforcement mechanism, until 1969. That year the U.N. General Assembly adopted a resolution condemning, as contrary to international law, use of any chemical or biological agents as weapons in armed conflict. It can be argued that this resolution codifies customary international law and hence takes precedence over the Geneva Protocol, making the prohibition against use of chemical weapons universal and eliminating the question of first use versus retaliatory use.[4] But it, too, left unresolved the questions of possession and military preparations and capabilities to use chemical weapons. The General Assembly can perhaps codify customary international law, and it clearly can establish an international norm of behavior, but any such articulation of international law can do no more than bolster a Security Council action finding a state to be a threat to the "international peace and security." Only a new treaty can unambiguously create a new obligation; without a treaty, states still had no basis for eliminating all chemical weapons in their arsenals, or for having confidence that others have done so.

In August 1968, the Conference of the Committee on Disarmament (CCD) placed the question of negotiation of a convention providing for chemical weapons disarmament on its agenda. As described in the previous chapter, the CCD took up the matter of more effective prohibitions against both chemical and biological weapons, and these discussions quickly led to a Biological Weapons Convention. Splitting the two issues, chemical and biological weapons, was explicitly a matter of convenience and an effort to gain rapid acceptance of a biological weapons prohibition.[5] In Article IX of the BWC, each state party "affirms the recognized objective of effective prohibition of chemical weapons" (possession, not just use) and "undertakes to continue negotiations in good faith with a view to reaching early agreement" on a chemical weapons convention.

While there were clear judgments in the international community that biological weapons were not militarily useful on the battlefield, chemical weapons were judged very differently. Many countries, including all the major powers, considered chemical weapons to possess great utility on the battlefield.[6] Whether openly acknowledged or not, most of the major powers had active offensive chemical weapons programs, as well as defenses against chemical weapons. The crux of the matter was not abolishing use, but abolishing possession. For this reason verification was a very important issue in the negotiations.

Those who negotiated the IAEA Statute, and even the NPT, were able to construct a control and inspection regime while the civilian scientific enterprise was in its infancy and the civil nuclear industry was largely still in prospect. The chemical weapons negotiators confronted a very different and significantly more difficult situation:

Over the last 1,000 years, the chemical industry developed as an integral part of modern society; a modern chemical industry is present to some degree in all of the significant negotiating states and economically vital for a great many of them; chemical weapons were widely believed to be in the arsenals of many participants, but few were willing to confirm this fact, either in public or behind closed doors. Under these circumstances, the negotiating process was certain to prove much longer and more complex.

Negotiating the Chemical Weapons Convention

From the beginning, it was clear that verification would be a central issue in the Chemical Weapons Convention (CWC) negotiations, notwithstanding its having been a lesser issue in the BWC and totally absent from the final agreement. The U.S. position was that the verification provisions in the convention would have to provide "adequate" assurance that other parties were complying. The Reagan administration defined adequacy to mean:

> The U.S. must have the ability to acquire sufficient information to render a reasonable judgment on whether other parties are complying with the limits of an agreement and its provisions, and to render this judgment in a timely manner, such that we can compensate for any risk posed to our security by the violation.[7]

There are a number of significant elements to this statement, not the least of which is the focus on the ability of the United States to make an independent decision regarding the compliance of others, in contrast to the IAEA/NPT situation. Of greater consequence for all parties, this standard identified three vital factors:

- Military significance as a detection threshold
- Timely warning of violations
- Timeliness being determined by the need for an effective response.[8]

The question for negotiators was what these criteria would mean in practice, and how to achieve that.

Early in the Reagan administration it was recognized that challenge inspections of some sort were the key to effective verification. However, the administration was split between those who desired a convention and those who opposed it because of the resultant restrictions on U.S. options. The latter were convinced that the Soviet Union was violating the BWC and would also violate any CWC. Their victory in the internal policy debate led to a U.S. position, enunciated by Vice President Bush in February 1983 before the CCD. Bush argued that the regime must include national declarations, systematic "international" onsite inspections of relevant facilities (which, by the character of the obligations contemplated, would involve many kinds of facilities), "a multilateral mechanism for dealing with compliance issues,"[9] and challenge inspections "anywhere, any time, with no right of refusal."[10] As had been expected, such intrusive inspections caused the USSR to reject the U.S. proposal promptly, and negotiations remained deadlocked for several years. The USSR remained adamant, as did the United States. The administration informed Congress in its fiscal 1986 Defense Department report,

> We realize that such a verification measure is unprecedented, but the risks of the status quo or of an unverifiable treaty are so severe that they far outweigh the risks of allowing international inspection teams into our sensitive facilities.[11]

As Michael Krepon observes, "cynicism and public relations constitute a hollow basis for government policy."[12] The United States had taken a position it, too, could not in fact accept. Many participants in the negotiations suspected as much, and it was confirmed when, in 1987, Mikhail Gorbachev reversed Soviet policy and accepted the U.S. proposal. The United States was now confronted with the prospect of having to accept its own proposal permitting international access to any facility in the United States, whether government, commercial, or private, thus jeopardizing the most sensitive national security programs, or backing away from its own proposal.

The British took on the task of seeking some more satisfactory approach to the challenge inspection problem and began to develop an approach labelled "managed access." In broad outline, under this approach a challenge inspection would be permitted "anywhere, any time" but it would not involve unfettered access. Rather, the inspected state would have rights to delimit access in certain respects. Inspectors would be permitted to perform those activities necessary to confirm that treaty violations were not being conducted at the inspected site but would not necessarily be able to determine what did in fact take place there.[13]

Over the next several years negotiations reduced the number of outstanding issues, and pressure mounted for some satisfactory resolution to the challenge inspection question. In July 1991 the United States put forward a new proposal, reflecting an evolution of the original British managed access concept. This proposal was co-sponsored by the United Kingdom, Australia, and Japan, but more out of loyalty than enthusiasm.[14] It proved, with some additional refinement, acceptable to the vast majority of negotiators.

While an embarrassing course of events for the Bush administration, this evolution of the inspection "anywhere, any time" issue did result in an innovative new balance between the conflicting requirements of effective verification and protecting national security activities, a problem confronting many governments in the CWC negotiations. In putting forward its initial "anywhere, any time, no right of refusal" proposal, the United States dug a hole for itself—but this hole had the valuable effect of focusing the debate on effective verification rather than on minimizing intrusiveness. The conceptual direction from which a compromise or balance is approached can make an important difference in the outcome.

The Conference on Disarmament completed its work in the autumn of 1992, and in January 1993 the CWC was opened for signature in Paris, at which time 130 countries signed. The Convention was to enter into force 180 days after the 65th ratification, "but in no case earlier than two years after its opening for signature."[15] The 2-year time period is necessary for the work of a Preparatory Commission, which is charged with establishing the Organization for the Prohibition of Chemical Weapons and developing the detailed inspection procedures and guidelines the organization will implement. President Clinton sent the CWC to the Senate in November 1993 for its advice and consent to ratification.

The Chemical Weapons Convention

The Convention on the Prohibition of the Development, Production, Stockpiling and Use of Chemical Weapons and on Their Destruction is arguably the most ambitious disarmament convention in history. Even its name indicates the wide scope of the undertakings. Each state party undertakes "never under any circumstances" to:

- Develop, produce, otherwise acquire, stockpile, or retain chemical weapons, and not to transfer chemical weapons "to anyone"
- Use chemical weapons
- Engage in any military preparations to use chemical weapons
- Assist, encourage, or induce, "anyone" to engage in any prohibited activity.

In addition, each party undertakes to destroy any chemical weapons it "owns or possesses," or that "it abandoned on the territory of another State Party," and to destroy "any chemical weapons production facilities it owns or possesses." Finally, each party undertakes not to use riot-control agents as a method of war.[16] This last undertaking had long been a contentious point concerning the Geneva Protocol.

These prohibitions impose substantial verification requirements on the entity created by the CWC to perform this function, the Organization for the Prohibition of Chemical Weapons (OPCW). Like the NPT and IAEA safeguards, the CWC charges the OPCW with responsibility for verifying some, but not all, of these undertakings. The OPCW is responsible for verifying that each state party:

- Does not henceforth produce or otherwise acquire, stockpile, or retain chemical weapons
- Destroys the chemical weapons in its possession at the time the CWC enters into force (or, for states joining later, is ratified)
- Destroys chemical weapons production facilities it owns or possesses.

In addition, the Director-General is responsible for investigating alleged use of chemical weapons by a party (if the alleged user is not a party, the Director-General "shall cooperate closely with the Secretary-General of the United Nations").[17] Verifying actual use of chemical weapons differs in many important respects from the other verification responsibilities given to the OPCW and is likely to be required only infrequently (while other verification responsibilities will be on-going, verification of use requires an accusation). The inclusion of use verification in the CWC is unique: the nuclear nonproliferation system does not address investigating alleged use (nor perhaps need it, use of a nuclear weapon being rather obvious), and the BWC relies on the authority of the U.N. Secretary General to create ad hoc groups of experts to conduct necessary investigations.

The CWC does not charge the OPCW with verifying the parties' behavior regarding military preparations (doctrine, training, or equipping) to use chemical weapons, or assistance or encouragement that the party might give to others to engage in prohibited activities. Verifying the absence of any secret training and equipping of military forces for offensive chemical warfare (defensive training and equipment being legitimate) would require a very different sort of inspection than any contemplated under the CWC, although challenge inspections may potentially detect such activities. Finally, verifying the absence of covert cooperation between states is most difficult and best addressed not by attempting to detect the cooperation, but rather by inspections to detect the illegal activity in the recipient state (assuming that it is party to the CWC).

Each of the three undertakings subject to routine verification (destruction of existing chemical weapons, destruction of production facilities, and no further production or possession) requires a somewhat different inspection approach, but draws largely from the same repertoire of verification methods. Destruction of existing chemical weapons and production facilities will be verified by each state first declaring all now-prohibited facilities and weapons. The state will also submit plans describing the destruction process to be implemented. Inspectors will then confirm these declarations, periodically inspect the destruction process to confirm progress according to the plans, and use equipment to monitor activities between on-site visits. During the first decade the CWC is in force, this will be a massive job for the OPCW, but as pre-existing weapons and related facilities are eliminated, this aspect of the verification effort will wind down.

Verifying that states are abiding by their commitments not "to develop, produce, otherwise acquire, stockpile or retain chemical weapons" entails two different verification activities:

Confirming that civilian chemical plants are producing only for civilian applications, even when producing chemicals which are immediate precursors to chemical weapons agents. This will be accomplished by regular receipt of inventory and transactions reports, audits of records, onsite inspections including taking samples and examination of design information and production activities, and perhaps the use of unattended instruments (e.g., time-lapse cameras) to monitor processes or areas.While many of the specific verification measures are the same as used by the IAEA for nuclear inspections, the actual verification approach will be quite different, focusing much more on activities and much less on establishing and confirming a material balance. Nuclear materials can in most instances be measured to the gram and the total quantities of material are such that this is practical, in addition to which the quantity of military significance justifies such precise measurements. Chemicals in a large commercial production facility cannot be measured with anywhere near comparable precision, nor is this justified by the quantities deemed to be of military significance. In controlling chemical weapons, hundreds to thousands of kilograms of chemical agent are judged a quantity of military significance,[18] while the nuclear material to fabricate a single nuclear explosive on the order of 10 kg of plutonium and maybe 3 times as much highly enriched uranium is the quantity deemed to be of military significance.[19]

The more difficult part of verifying compliance with commitments not "to develop, produce, otherwise acquire, stockpile or retain chemical weapons" comes with the question of hidden stockpiles and covert production facilities. As noted in chapter 2, this aspect of the nuclear verification problem defied solution from the time the NPT was negotiated until very recently and even today remains to be fully and convincingly implemented. The CWC attempts a new approach: challenge inspections involving "managed access." The problem has been addressed before,[20] but the problem is always how to balance intrusiveness against credibility of the result. Managed access challenge inspections may represent the answer, but equally important are the steps necessary before a challenge inspection takes place.

Institutions of the CWC

Implementing such responsibilities requires both institutions and procedures. The CWC establishes both and, by comparison with the NPT, does so in great detail, especially with respect to the procedures. In addition, the treaty creates three institutions: the Technical Secretariat, the Executive Council, and the Conference. These issues are treated again in an Annex on Verification and Implementation, which runs more than twice the length of the treaty itself.

The first of these is the Technical Secretariat, the international civil service responsible for performing, among other things, the day-to-day work of verification. The Technical Secretariat "shall carry out the verification measures provided for in this Convention"[21] (the significance of "measures" rather than "responsibilities" will be discussed below). The Technical Secretariat is to inform the Executive Council of:

> any problem that has arisen with regard to the discharge of its function, including doubts, ambiguities or uncertainties about compliance with this Convention that may have come to its notice in the performance of its verification activities and that it has been unable to resolve or clarify through its consultations with the State Party concerned.[22]

Intriguingly, the CWC goes so far as to stipulate that all inspectors and other staff of the Technical Secretariat must be citizens of a party,[23] in addition to the expected requirement that they "shall not seek or receive instructions from any Government or any other source external to the Organization."[24]

The Executive Council is to be the "executive organ of the Organization" and "shall promote the effective implementation of, and compliance with, this Convention."[25] It will supervise the activities of the Technical Secretariat, prepare the annual program and budget for Conference action, and otherwise be responsible for relations between the Organization and the member states.[26] The Executive Council is given broad powers to:

> consider any issue or matter within its competence affecting this Convention and its implementation, including concerns regarding compliance, and cases of non-compliance, and, as appropriate, inform States Parties and bring the issue or matter to the attention of the Conference.[27]

When confronted with "doubts or concerns regarding compliance and cases of non-compliance," the Executive Council is to consult with and request the noncomplying party to redress the situation "within a specified time." If further action is considered necessary, the Council may (a) inform all parties, (b) inform the Conference, or (c) recommend to the Conference measures to ensure compliance. In cases of "particular gravity and urgency" the Executive Council may take the matter directly to the United Nations General Assembly and Security Council.[28]

These provisions are formulated very carefully, and reflect a difference of view between the United States and most of the other states which participated in the negotiations. Apparently most states in the negotiation hold the view that these responsibilities parallel rather precisely those assigned by the Statute to the IAEA Secretariat and Board of Governors. The United States understands them differently.[29] The IAEA Secretariat determines whether or not a state has violated a safeguards agreement (as opposed to having violated the NPT, or even the Statute itself), and the Board of Governors determines what course of action to pursue, acting on a recommendation from the Director General.[30] Except for the case of challenge inspections, the responsibilities of the Executive Council with respect to noncompliance are to address "what next." Neither the Executive Council nor the Technical Secretariat is explicitly given the authority to determine that a violation of the CWC has been identified. In the U.S. view, the Technical Secretariat will simply inform the Council of ambiguities or doubts (in the words used by the IAEA on several occasions, that it is "unable to verify compliance").

During the CWC negotiations, the United States took the position that only individual member states can judge whether or not another state is complying with its obligations, and the lack of explicit authority for either the Technical Secretariat or the Executive Council to judge compliance is the result. This absence of clearcut authority to judge compliance could in the future create some thorny legal issues, such as even when there is no ambiguity and consensus exists within the Executive Council that a violation has been detected, must it nonetheless report the matter to the Security Council as simply an unresolved problem, or what are the powers of the Security Council (or a Review Conference of CWC parties) to judge compliance and noncompliance?

In practice the vast majority of cases probably will not constitute clearcut cases of noncompliance but rather situations in which the compliance of a party is subject to significant doubt and remedial action is required to remove the doubt. Given its behavior with respect to IAEA safeguards, one can imagine how North Korea might use such a situation to defend its failure to comply. The authority to declare a state in noncompliance is a powerful diplomatic tool in any conflict of wills. Lacking this authority, the ability of the Executive Council and the Conference to haul a violator before the court of international opinion is substantially reduced.

Although neither the Executive Council nor the Conference is specifically empowered to judge compliance, the Conference has substantial authority to pursue any failure by a state party to rectify noncompliance. Not only is the Conference to "take the necessary measures to ensure compliance with this Convention and to redress and remedy any situation which contravenes the provisions this Convention," it can, upon recommendation of the Executive Council, "restrict or suspend the State Party's rights and privileges" until it conforms to its obligations.[31] Furthermore, in cases where "serious damage to the object and purpose of this Convention may result from activities prohibited under the Convention," the Conference may even "recommend collective measures to States Parties in conformity with international law,"[32] which would include the United Nations Charter. The Conference "shall in cases of particular gravity, bring the issue, including relevant information and conclusions, to the attention of the" United Nations General Assembly and Security Council.[33] There is some irony that, on the one hand, the CWC grants to the Conference powers of collective action, which would appear to include military action, while on the other hand cases of "particular gravity" are to be referred to the Security Council.

Unlike previous arms control and nonproliferation treaties, the CWC grants its own institution the power to mandate force against a violator, but withholds the authority to make a specific finding of noncompliance. Given the central role of the United States in negotiating the CWC, the U.S. position appears to reflect a balance between two concerns. The United States worried that the CWC Executive Council would not declare noncompliance in some case, but the President wanted to report a treaty violation to Congress.[34] In effect this passes the matter of international judgment regarding noncompliance up to the Security Council, where the veto can be exercised. It also means that a decision regarding the need for collective action against a violator can be handled in the CWC Conference, where a majority can make the determination without any veto.[35]

Verification Annex

The Annex on Implementation and Verification, a substantial document bearing much the same relationship to the CWC as INFCIRC/153 bears to the NPT, has responsibility to:

- Establishes rights and obligations of the Technical Secretariat and states parties relevant to the conduct of verification activities
- Identifies specific procedures and means for accomplishing verification activities
- Establishes what might be called technical policies.

With respect to the first, the CWC is intended to be a self-executing treaty, without the need for separate bilateral agreements between the Organization and individual states. This is in contrast to the need for a separate safeguards agreements with the IAEA to implement NPT verification obligations. However, "facility agreements concluded between States Parties and the Organization"[36] are specifically called for. These are to be similar to the "subsidiary arrangements" and "facility attachments" used in NPT safeguards to translate general methods and practices into terms appropriate for a particular country (its legal and administrative context) and for individual facilities (every plant is a little different, and some are very different).

The approach and level of detail reflected in the Annex's treatment of procedures and technical policies are very specific; at the same time, it sometimes elevates second-order technical issues to political levels of decisionmaking. One example: the Annex stipulates that inspection equipment "shall . . . be designated, calibrated and approved by the Technical Secretariat," which it clearly should be, but then goes on to require that "Designated and approved equipment shall be specifically protected against unauthorized alteration."[37] Whether in every case tamper-proofing is practical, or even technically feasible, much less necessary, is not addressed.

Such matters of excessive detail can be worked out in the first few years of implementation, but the Annex also creates what may prove to be a structural problem. It provides a substantial basis for micromanagement of technical matters by the Conference:

> Detailed Procedures for the conduct of inspections shall be developed for inclusion in the inspection manual by the Technical Secretariat, taking into account guidelines to be considered and approved by the Conference pursuant to Article VIII, paragraph 21(i).[38]

The referenced paragraph simply states that the Conference may "consider and approve at its first session any draft agreements, provisions and guidelines developed by the Preparatory Commission." Whether Conference approval of all such materials can be accomplished at one session is unknown, but it may result in the politicization of highly technical matters bearing directly on whether onsite inspections are credible and technically effective. Although responsibility for determining technical policies has a long history of controversy in the IAEA, the ability of the IAEA Secretariat to determine technical policies largely on its own has been essential to maintaining technical credibility and facilitating improvements in methods. How this issue will play out in the chemical weapons context will become known only over the course of implementation, but it could permit technical matters to become counters in political controversies. The objective was political ratification of technical issues, and if this is the end result it will be positive. If the result is substantive changes being made in a political forum for political reasons, however, it is unlikely to improve either the effectiveness or the credibility of the verification system.

The Verification Annex gives OPCW inspectors a number of important rights beyond those historically afforded IAEA inspectors (the IAEA is now trying to retrofit some of these rights). In some respects OPCW inspectors will have rights more like those afforded to the U.N. Special Commission and the IAEA by Security Council Resolution 687 for inspections in Iraq. For example, inspectors have the right to use their own telecommunications equipment while onsite, for communicating both within the team and with headquarters,[39] and in addition, communications with headquarters may be encoded.[40] Obtaining visas has been a constant problem for IAEA inspectors, as many states require a separate visa for each entry, which sometimes delays inspections and always signals the state and facility of an impending inspection well in advance. The Annex stipulates that OPCW inspectors will receive multiple-entry visas of 2 years duration from each relevant state.[41] Inspectors will be required to use designated points of entry, but these must permit reaching "any inspection site" within 12 hours.[42] (Oddly, in the case of challenge inspections, the requesting party, not the inspected party or the inspectorate, designates the point of entry to be used.)[43] A major point of contention between the United Nations and Iraq has been transportation of inspectors within Iraq and particularly the use of aircraft (as issue never even contemplated for regular IAEA inspections). The Annex stipulates that OPCW inspectors may use organization-owned or chartered aircraft, and the state is obligated to give "standing diplomatic clearance" for such aircraft.[44]

The IAEA has frequently had difficulty gaining acceptance for new methods and new instruments (facility operators commonly raise safety questions), even though each safeguards agreement stipulates that the IAEA is to "take full account of technological developments in the field of safeguards.[45] OPCW inspectors should not confront the same difficulties, as the Annex stipulates that "there shall be no restriction by the inspected State party on the inspection team bringing onto the inspection site such equipment, . . . which the Technical Secretariat has determined to be necessary to fulfill the inspection requirements."[46] However, procedures and equipment regulations in different countries are frequently in conflict, and the

IAEA has experienced a variety of such problems with inspection equipment (most of which can be worked out with good will, but provide opportunities for those inclined to make matters difficult either for political reasons or by virtue of personality). The Preparatory Commission is developing the list of equipment to be approved for inspection use. As technologies develop, some equipment will become obsolete and new equipment will become available; how equipment will be added to and removed from the list is unclear.

While the IAEA Statute gives inspectors the right to meet with and question "any person who by reason of his occupation deals with [items] which are required by this Statute to be safeguarded,"[47] this right has not been enforced, or perhaps even attempted since the first IAEA inspections. In addition, there may be legal questions as to whether it applies to NPT safeguards.[48] No such doubts should exist for OPCW inspectors, as they are clearly given the right to interview any facility personnel, albeit in the presence of inspected state representatives.[49] At the same time, the state representative may object to questions on germaneness grounds.[50] For this provision to prove useful inspectors will have to use it from the beginning, and Technical Secretariat management and the Executive Council will have to support its enforcement.

The Verification Annex also identifies certain limitations on inspectors, which, for the most part, will not be significant unless inappropriately enforced by states and the Executive Council fails to support the Technical Secretariat in a reasonable interpretation (or unless the Director-General proves reticent even to bring such matters to the Council).

After completing an inspection, the inspection team "shall meet with representatives of the inspected State Party . . . to review the preliminary findings of the inspection team and to clarify any ambiguities." The inspected state will also receive the inspection team's preliminary findings in writing.[51] Just what is meant by the inspection team's "preliminary findings" can only evolve through practice, but inspectors may face pressure to give a "clean bill of health" before departing the country or to specify precisely the additional inspection activities or information needed to permit them to do so. The requirement that "verification activities . . . shall only be performed by designated inspectors" may form the legal basis for arguments that headquarters personnel, including lawyers, inspectorate management, and the Director-General, are not to be consulted before such a determination is made.[52] As noted above, such problems can be resolved only by a diligent inspectorate and a Council with the will to support the inspectorate.

Challenge Inspections

As just implied, the conflict between verification and national sovereignty has always been the principal impediment to effective verification measures. It was as difficult an issue in negotiating the CWC as in previous negotiations, but the product of these most recent negotiations is very different and a substantial advance for verification rights over sovereignty rights. Whether the provisions of the CWC are rigorously implemented, or whether like many provisions of IAEA safeguards, they are severely curtailed in practice, the frontier of international inspection rights over sovereignty rights has been advanced. Nowhere is this better demonstrated than in the matter of investigating compliance questions raised by one party with respect to another.

Each party has the right to request clarification of the activities of any other party. When the OPCW receives such a request, the first step is for the Executive Council (not the Technical Secretariat) to provide to the requesting state that information in its possession relevant to the compliance of the challenged state.[53] The concerned state party may also ask the Council to request clarification from the challenged state. The Council has 24 hours to act on this request, and the accused state has 10 days to respond.[54] If these efforts fail to resolve the matter to the satisfaction of the requesting state, it can request that the Executive Council have the Director-General establish a panel of experts (which may include both staff and outside personnel) to review the information and provide a factual report, which goes both to the Council and to the state expressing concerns.

In the event that the concerned state is still not satisfied, it may request a challenge inspection of the "accused."[55] The Technical Secretariat is charged with performing the challenge inspection, but the challenging state may send an observer.[56] The inspected state need not grant the observer the full access afforded the inspectors.

The Director-General is given strict requirements for prosecuting the challenge inspection.

The Executive Council may review the challenge inspection request before it is acted upon, determine that it is inappropriate, and stop it.[57] However, the Council must act within 12 *hours* of receiving the challenge inspection request and must decide by a three-fourths majority that the request is "frivolous, abusive or clearly beyond the scope of this Convention."[58] As most smaller countries do not have diplomatic missions resident in The Hague, it is highly unlikely that the Executive Council will be able to convene, much less act to block, a challenge inspection.[59]

Each party may request an onsite challenge inspection of "any facility or location in the territory or in any other place under the jurisdiction" of any other party and to have this inspection conducted "without delay."[60] But the requesting state is obligated "to keep the inspection request within the scope of this Convention and to provide in the inspection request all appropriate information on the basis of which a concern has arisen."[61] A challenge inspection gives unprecedented access for inspectors,[62] who are to gain access to the site "as soon as possible, but in any case not later than 108 hours after" entering the country.[63] Inspectors may perform aerial observation of the site,[64] have a right to take swipe, air, soil, and effluent samples around the site perimeter,[65] and may (subject to negotiation of specific sampling procedures) take samples within the site.[66] With the exceptions noted below, inspectors are to have access to all places and buildings within the designated perimeter.[67]

While the "accused" state must accept the challenge inspection, it may take a number of steps to limit the scope of the inspection, to "manage access." The requesting party specifies the perimeter of the site to be inspected (according to rules identified in the Annex), but the inspected party may, within 24 hours of the inspectors' arrival, propose an alternative perimeter, provided that it "include[s] the whole of the requested perimeter."[68] The inspection team is to have access to all vehicular traffic exiting the site and to all locations within the identified perimeter within 12 hours of arriving at the site.[69] The inspected state may, in exceptional cases, give only individual inspectors access to certain parts of the inspection site.[70] Access to vehicular traffic and certain portions of buildings can be limited, but the inspected state "shall make every reasonable effort to demonstrate to the inspection team" that the vehicle or area to which access is denied "is not being used for purposed related to the possible non-compliance concerns raised in the inspection request."[71] When granting access, the inspected state may remove sensitive papers, shroud sensitive items, turn off computers and "data indicating devices," restrict sample analysis to detecting the presence or absence of chemicals listed in CWC Schedules and require that the analysis be performed on-site, and require use of random sampling techniques for access to items and certain areas.[72]

In sum, the inspected state has a variety of tools to manage access. It can provide information it believes will demonstrate compliance, obviating the need for access. With the agreement of the inspectors, the inspected state can delimit the inspected area in certain respects. The inspected state can shroud equipment and activities "to prevent disclosure of confidential information and data, not related to this Convention." This "managed access" approach permits what amounts to "anywhere, any time" inspections while ensuring that sensitive facilities and activities not related to CWC objectives can be properly protected.

Following the inspection, the Executive Council has a second chance to consider issues of potential abuse of the challenge inspection right, when it reviews the final report of the inspection team. Specifically, the Council is charged to address any concerns as to whether:

- Any non-compliance has occurred
- The request had been within the scope of this Convention
- The right to request a challenge inspection had been abused.[73]

This explicit authority to determine whether compliance has been confirmed or whether violation has been detected stands in contrast to the situation for routine inspections (described above). The Council is also to decide whether further corrective action is needed, either to ensure compliance by the inspected state, or "in the case of abuse, whether the requesting State Party should bear any of the financial implications of the challenge inspection."[74]

In as much as the Executive Council is to "address any concerns as to whether any non-compliance has occurred" on the basis of reviewing the "final report of the inspection team," this situation appears similar to that described in the IAEA Statute, where "the inspectors shall report any non-compliance to the Director General." In practice it is not IAEA inspectors who determine noncompliance, but the IAEA Director General acting on the recommendation of the corporate body—the inspectors and their management, headquarters personnel responsible for laboratory analysis and data evaluation, and the legal and diplomatic offices. The situation will certainly be much more politically charged in the case of a CWC challenge inspection, and for this reason the OPCW Director-General may hesitate to insert his own judgment in the form of a recommendation to the Council. Nonetheless, there will also be pressures from members of the Council for a technical recommendation as opposed to simply providing the relevant data: interpreting technical data requires technical, not political, capabilities and judgments. The first few hard cases will determine practice, and a close reading of the text and negotiating history will be balanced against the political requirements of the situation.

Next Steps

The treaty requires at least 2 years between its initial signing and entry into force, in order to permit the Preparatory Commission to complete its work. Depending on the speed with which states ratify, it may be longer, but it cannot be shorter. The Commission was established by a special resolution of the signatories "for the purpose of carrying out the necessary preparations for the effective implementation" of the Convention.[75]

In fact, the Preparatory Commission has the responsibility of filling in most of the detail not specified in the Convention itself. This includes "elaboration of a detailed staffing pattern of the Technical Secretariat," which means determining staffing requirements, staff rules, procurement procedures, and all the other administrative minutiae necessary to have a functioning organization. The Commission is required to:

- Prepare a draft program and budget for the organization's first year

- Develop arrangements for the election of the first Executive Council
- Draft verification guidelines as identified throughout the Verification Annex
- Prepare an initial list of approved inspection equipment; draft safety procedures for inspection activities
- Develop model facility agreements.

In essence, the Preparatory Commission is responsible for preparing all the rules, guidelines, procedures, and other documentation necessary for the OPCW to open its doors and begin functioning as a mature operating agency, to be in its first year what the IAEA has become for nuclear safeguards in its 30th year. To maintain the credibility of the CWC in the face of the many challenges it must confront, it cannot begin slowly or falter significantly. This is a wise effort, but it is also incredibly ambitious.

Analysis

From a verification perspective, the Chemical Weapons Convention is the most ambitious disarmament treaty every attempted. It seeks to eliminate universally an entire category of weapons. This requires verifying that all declared chemical weapons programs (in states party to the CWC) are in fact entirely dismantled and all existing stocks of weapons and agents are destroyed. At the same time, it means checking to ensure that one of the largest and most widespread industries in the world is not, at any place, misused to fabricate prohibited chemical warfare agents. It entails establishing the most intrusive inspection system ever and operating that system in such a manner that all participating governments believe that any militarily significant violation would be detected.

The negotiators of the CWC sought to understand the lessons from NPT safeguards and reflect these lessons in their product. The governments of at least several major states participating in the negotiations conducted studies and symposia to identify positive examples, problems, and poor approaches in the IAEA/NPT model. Thus the routine inspection regime negotiated into the CWC reflects models taken from the IAEA/NPT experience; solutions to problems experienced in the IAEA/NPT case; and new approaches tailored to the different situation presented by the chemical industry and by the need to verify elimination of chemical weapons (which has no analogue in IAEA experience).

The challenge inspection system, with managed access to all places in every state party, must be implemented not only to establish credibility for detecting violations, but also to maintain the view that the inspection system itself is a benefit rather than a threat to national security. If either governments or the commercial chemical industry in significant countries come to believe that the intrusiveness of challenge inspections costs more than the security benefits, the treaty itself may come under attack when the OPCW seeks to implement it and in the first review conference. One might argue that in part the NPT survived its first 25 years more because inspections did not threaten national values than because the treaty provided the security guarantees originally contemplated, and is only now becoming a truly effective nonproliferation treaty. The CWC cannot afford a similarly slow growth.

Notes

1. U.S. Arms Control and Disarmament Agency [ACDA], *Arms Control and Disarmament Agreements*,(Washington, DC: U. S. Government Printing Office, 1990), 3-4; see also the 1899 Hague Declaration 2 Concerning Asphyxiating Gases and article 23 of the Annex to the 1907 Hague Convention IV Respecting the Laws and Customs of War on Land, both reproduced in Adam Roberts and Richard Guelff (eds.) *Documents on the Laws of War*, second edition (Oxford: Clarendon Press, 1989).
2. ACDA, 10.
3. Nicholas A. Sims, *The Diplomacy of Biological Disarmament Vicissitudes of a Treaty in Force, 1975-85* (New York: St. Martin's Press, 1988), 38-42.
4. The general view of legal scholars is that the General Assembly cannot create international law, and hence the import of a General Assembly resolution taking note of customary international law is likely to be determined by the degree to which there is a consensus regarding the judgment that such was customary international law prior to the General Assembly resolution.
5. Sims, 34.

6. See Victor Utgoff and Susan Leibbrandt, "On the Pursuit of Universal Adherence to the Chemical Weapons Convention," in *Chemical Disarmament and U.S. Security*, ed. Brad Roberts (Boulder: Westview Press and Center for Strategic and International Studies, 1992), 36-38.

7. Quoted in Michael Krepon, "Verification of a Chemical Weapons Convention," in Roberts, footnote 3, 88.

8. Krepon, 73.

9. U.S. Arms Control and Disarmament Agency, *Fact Sheet, Chemical Weapons Negotiations at the Conference on Disarmament*, August 13, 1992.

10. Ronald F. Lehman, "Concluding the Chemical Weapons Convention," in Roberts, 8.

11. Quoted in Krepon, 82.

12. Krepon, 82.

13. See Krepon, for a more detailed discussion of the internal U.S. debates and the development of the "managed access" concept.

14. Krepon, 84.

15. Convention on the Prohibition of the Development, Production, Stockpiling and Use of Chemical Weapons and Their Destruction, Paris, January 13, 1993 ["CWC"], (reprinted by ACDA); Article XXI.

16. CWC, Article I.

17. CWC, Verification Annex, Part XI, "Investigations in Cases of Alleged Use of Chemical Weapons." Quotation from paragraph 27.

18. See Victor Utgoff and Susan Leibbrandt, 36-38, for the best discussion of this issue. Their numbers are supported by many others writing on the subject, but without the same detailed analysis to support those numbers.

19. A single (relatively small) nuclear weapon can easily kill thousands to tens of thousands of people, whether used to attack a city or military formations. This is one reason the detection thresholds for nuclear materials are set by the quantities of highly enriched uranium or plutonium necessary to fabricate a single nuclear explosive. It appears that the same scale of casualties (but only by orders of magnitude) is used to determine the quantities of chemicals judged to be militarily significant.

20. IAEA safeguards implementing the NPT involve "special inspections," which (like a search warrant in the United States) require some form of probable cause beyond mere suspicion. The Treaty of Tlatelolco ("Treaty for the Prohibition of Nuclear Weapons in Latin America") provides for challenge inspections when one State so requests of another, but this has never been done. The Treaty originally provided roles for both the IAEA and OPANAL (the Agency for the Prohibition of Nuclear Weapons in Latin America) in challenge inspections. In 1993 the Treaty was amended at the

behest of Argentina and Brazil to limit OPANAL's role, but not that of the IAEA, in these inspections.

21. CWC, Article VIII/37.

22. Ibid., VIII/40.

23. Ibid., VIII/44.

24. Ibid., VIII/46. The same provision is contained in the IAEA Statute. Different member states have treated this provision very differently. The United States has honored it quite rigorously for the most part, while some states have considered their nationals employed by the Secretariat to be literally representatives responsible as much for protecting the national interest as to conducting the work of the Secretariat. In recent years it appears that the trend is towards more independence for all Secretariat staff, but national politics is a permanent feature of any international organization (for people from different countries bring different perspectives and values, even when they do not seek or receive instructions from home).

25. Ibid., VIII/30 and VIII/31, respectively.

26. Ibid., VIII/31 & 32.

27. Ibid., VIII/35.

28. Ibid., VIII/36. For some reason, perhaps reflecting the interests of the majority of parties which will not be members of the Security Council at any one time, not only is the General Assembly always referred to, but it is always referred to first.

29. Interviews with State Department (November 5, 1993), ACDA (November 5, 1993; December 8, 1993; and February 18, 1994), and Defense Department (November 24, 1993) officials.

30. Although the language of Article XII.C of the IAEA Statute gives this responsibility directly to the inspectors, in the next sentence it gives this same responsibility to determine non-compliance directly to the Board of Governors. In either case, it is the IAEA (rather than member states) which has the authority to determine non-compliance with safeguards agreements, which are with it. As explained in chapter 2, determining non-compliance with the NPT is a more ambiguous question, and the authority appears to reside with the Security Council.

31. CWC, Article XII.

32. Ibid., XII/3.

33. Ibid., XII/4. In as much as the Conference itself has the authority to recommend "collective action" against a violator, it seems that Chapter VII of the U.N. Charter has in some sense been imported directly into the CWC. Reference to the United Nations thus continues to have great political significance, but is no longer the only mechanism for employing force against a violating state.

34. Interviews (see note 29).

35. That is, the result is to codify in the CWC the same result sought by the United States in the 1950 General Assembly's "Uniting for Peace" Resolution (Resolution 377A (V) of the General Assembly, 3 November 1950). That is, make sure that the veto power which permanent members of the Security Council possess cannot be used to prevent a decision to take collective action when the general membership of the organization deems it necessary.

36. CWC, Annex on Implementation and Verification ["Verification Annex"], Section II/38.

37. CWC, Verification Annex, II/28.

38. Ibid., II/42.

39. Ibid., II/44.

40. Ibid., II/4.

41. Ibid., II/10.

42. Ibid., II/16.

43. Ibid., X/B/4(b).

44. Ibid., II/22.

45. INFCIRC/153, para. 6.

46. CWC, Verification Annex, II/27.

47. Statute of the International Atomic Agency, Article XII.A.6.

48. One could construct a legal argument that the safeguards regime established in Article XII of the Statute applies to items or activities required to be safeguarded by the Statute, but does not apply to other safeguarded agreements between the Agency and a member state entered into pursuant to the second clause of Article III.A.5.

49. CWC, Verification Annex, II/46.

50. Ibid.

51. Ibid., II/60.

52. Ibid., II/3.

53. This terminology (referring to the accused state rather than the "accused" state) appears consistent with the nature of the challenge inspection process. Unless the requesting state is willing to make what amounts to an accusation, there is no basis for proceeding with the process.

54. CWC, Article IX.

55. Ibid., IX/8.

56. Ibid., IX/12.

57. Ibid., IX/17.

58. Ibid., IX/17.

59. Interview with ACDA official, February 18, 1994.

60. CWC, Article IX/8.

61. Ibid., IX/9.

62. Unprecedented for any agreement states have entered into freely, access is not quite as complete as that afforded UNSCOM and IAEA inspectors acting under Security Council Resolution 687, which was imposed on Iraq following its 1991 defeat in the Gulf War.

63. Ibid., X/39.

64. Ibid., X/40.

65. Ibid., X/36.

66. Ibid., X/47 & 48.

67. Ibid., X/38 & 41.

68. Ibid., X/17.

69. Ibid., X/23-31.

70. Ibid., X/48(g).

71. Ibid., X/29, for "places, activities, or information, see Annex, X/42.

72. CWC, Verification Annex, X/48.

73. CWC, Article IX/22.

74. Ibid., IX/23.

75. "Resolution Establishing the Preparatory Commission for the Organization for the Prohibition of Chemical Weapons," by the states signatory to the CWC, At Geneva, 3 September 1992.

5. ❖ Limiting Conventional Forces in Europe

In the early 1980s, the failure of the Mutual Balanced Force Reductions (MBFR) talks was evident to all, but the issue of how to ease tensions and reduce conventional military force deployments in central Europe remained. These same issues had been the subject of other talks, beginning in 1972, conducted under the aegis of the Conference on Confidence- and Security-Building in Europe (CSCE). Where the MBFR talks had approached the issue from the perspective of defining and then reducing troop strength, the CSCE approach sought to identify measures that, if implemented, would increase the confidence of each side as to the nonaggressive intentions of the other, and to proceed with implementation of such measures as mutually agreed political acts rather than as treaty commitments.[1]

The first stage in identifying and implementing mutually acceptable confidence-building measures (also known as confidence- and security-building measures, or CSBMs) began with agreement on the Concluding Document of the Conference, which was signed in Helsinki in August 1975. The Helsinki Final Act provided for advanced notice of military maneuvers involving more than 25,000 troops taking place anywhere in Europe (all European countries except Albania were CSCE signatories) or in adjacent waters. Helsinki also established the possibility for each side to observe maneuvers conducted by the other, not as a right but when mutually acceptable. These measures were an important first step toward reducing the "dangers of armed conflict and of misunderstanding or miscalculation of military activities which could give rise to apprehension."[2] Because some signatories were not members of either alliance, the provisions of the Final Act address what individual states are to do and do not refer to either the North Atlantic Treaty Organization or the Warsaw Treaty Organization.

In 1984, at one of the CSCE meetings called for in the Helsinki Final Act, agreement was reached on convening a special meeting to "undertake in stages, new, effective and concrete actions designed to make progress in strengthening confidence and security and in achieving disarmament, so as to give effect and expression to the duty of states to refrain from the threat or use of force in their mutual relations."[3] This rather modest goal was to be pursued for 2 years and 9 months in Stockholm in the Conference on Confidence- and Security-Building Measures and Disarmament in Europe (known as the CDE).

The first 18 months of the CDE were devoted to discussing whether the product should be primarily declaratory or should identify concrete measures to be implemented. Another year was spent identifying such concrete measures, but no progress was made on the question of compliance. It was not until the summer of 1986, when the Soviet Union changed its long-standing policy and agreed to accept onsite inspections, that a credible outcome became possible.[4] (It was September 1986 when Soviet Ambassador Issraelyan "startled" the BWC review conference by proposing negotiation of a supplementary protocol to create verification procedures for that treaty.)[5]

Given the utterly incredible title of "Document of the Stockholm Conference on Confidence- and Security-Building Measures and Disarmament in Europe Convened in Accordance with the Relevant Provisions of the Concluding Document of the Madrid Meeting of the Conference on Security and Co-operation in Europe," the CDE final document was, like its predecessors, a political agreement rather than a legal (i.e., treaty) agreement. However, it provided for important new confidence-building measures. The most important single aspect of this agreement was the provision for onsite inspections to verify compliance. While the Helsinki Final Act had included onsite observation of maneuvers as one of the confidence-building measures, a signatory was not required to accept any particular request for observation. The CDE agreement, signed in the late summer of 1986, made it mandatory that observers from all other signatories be invited. The agreement also provided for onsite inspections to verify compliance with the other CSBMs in the agreement. Inspections were to be performed by each signatory of others as it might choose, and signatories could not refuse inspections unless the requests received exceeded an annual threshold.[6] These observation and inspection features signalled a substantial increase in the confidence each signatory could draw from the new agreement; instead of simply being provided with formerly sensitive or classified information by adversaries, it was now possible to actually go and collect one's own information confirming those reports.

At the next meeting of the CSCE, in Vienna in January 1989, it was agreed to convene a further special meeting to resume the negotiations on CSBMs. However, during 1988, NATO and Warsaw Treaty Organization members began discussions on a treaty to limit conventional forces in Europe, and a negotiating mandate was agreed to in January 1989. These latter negotiations produce the Conventional Forces in Europe Treaty.

Negotiating Conventional Force Reductions

Negotiations, which ran from March 1989 to November 1990, would prove to be historic, taking place during one of the most momentous periods in modern European history. The Western alliance entered these negotiations with three objectives:

- To establish a "secure and stable balance of conventional forces at lower levels"
- To eliminate "disparities prejudicial to stability and security"
- To eliminate the Warsaw Pact's capability to launch a surprise (short warning) attack.

In the judgment of at least one participant, these objectives would be secured,[7] and the realization of at least a third of these objectives was due as much to external events as to the negotiations.

Unlike previous CSCE agreements, this one was explicitly limited, to be negotiated between members of NATO and the Warsaw Treaty Organization, at least in conception. However, as the negotiations proceeded, the political context for the negotiations changed fundamentally. The Warsaw Pact dissolved, five parties changed their names and forms of government, and one party (the German Democratic Republic) disappeared.[8] For many former members of the Warsaw Pact, the purpose changed from reaching accommodation with the West to obtaining some control over the size and deployment of Soviet forces, albeit without the knowledge that shortly these would become Russian forces instead.

To date most of the public discussion of these negotiations, whether by participants or outside observers, has focused on the reductions in conventional forces and the limitations on deployment which were achieved. Relatively little attention has been paid to the verification elements of the agreement. No doubt this reflects the fact that the limitations in size and deployment of conventional forces in Europe represent significant achievements, while the verification procedures incorporated in the agreement are conceptually simple and based on previous CSCE confidence-building measures. However, from the perspective of this study, there are several interesting aspects to the verification procedures and their development.

First is the question of how to verify such force limitations. As the negotiations opened the United States government, if not all NATO participants, held the conviction that verification was politically necessary, whether or not anyone knew how.[9] The example of the Washington Naval Conference of 1922 stood vividly in the minds of many. The dreadnought (battleship) limitations achieved in that agreement were evaded by building battle cruisers and aircraft carriers. Fungibility among different types of conventional forces would present major problems both for the limits themselves and for any verification mechanism. The MBFR talks had sought to avoid the fungibility problem by measuring military capability in terms of troop strength but had failed over the problem of how to confirm declarations (and concerns among NATO allies that the USSR was understating its force strength even in the negotiations). In the new negotiations it was therefore essential to establish limits on some group of weapons systems that collectively provide an effective measure of conventional military capability, but at the same time define "end-item concentrations" in a fashion amenable to determination by national technical means.[10]

Over the course of the negotiations, solutions were found. Weapons to be limited by the treaty would, with minor exceptions, be large and distinctive enough to permit detection by national technical means, thus the limits on tanks, armored personnel carriers, and rotary and fixed-wing aircraft. As defined, "artillery" is less amenable to verification by national technical means, but this is the price for ensuring that treaty limits could not be evaded as were those of the 1922 treaty (by building less capable but still effective weapons systems, in this case slightly smaller caliber artillery). The question of what quantity of each item was militarily significant was also important. All participants acknowledged that violating the treaty limits by dozens of tanks or aircraft would not make a crucial difference, thus making detection thresholds fairly robust.[11]

The question of what entities would perform verification also changed during the negotiations. Early in the negotiations the NATO parties tabled a draft text in which inspections would be performed by each treaty organization (NATO and WTO).[12] However, as the negotiations proceeded, members of the Warsaw Pact began to think in independent terms, and then the Pact dissolved. As this process took place, many of the Eastern participants began to think in terms of inspecting the Soviet Union as much as of inspecting NATO countries. The original NATO proposal was replaced by a new proposal in which each state party to the treaty would be permitted to inspect any other. While force limits continued to be cast in terms of two groups of states (reflecting NATO and the former Warsaw Pact alliances), actual force-size obligations and all other treaty obligations inhere to individual states parties, not to alliance organizations.

The treaty was signed November 19, 1990, as changes in Europe continued,[13] but the most dramatic change was yet to come and would delay entry into force for 2 years. On December 31, 1990, the Soviet Union dissolved, creating eight parties instead of one. This had several consequences, among them the need for an understanding as to how the forces previously allocated to the USSR were to be allocated among the new parties. The allocation question was both facilitated and complicated by limitations on where forces could be deployed. The CFE treaty defines zones within which permitted forces must be located, and these zones were defined in terms of the previous east-west alignment expecting conflict at the Fulda Gap. The original idea was to move forces away from the Fulda Gap, with little attention being given to the north-south dimension.[14] However, the defined zones did severely limit the deployment of forces along the "flanks." In the new geopolitical order, it is precisely these flanks that are of greatest interest and concern to Russia and many other parties. Hence there is now great pressure to revisit the questions of "flank limits" and zone structure and equally great pressure by others to maintain the status quo.[15]

Even though the demise of the Soviet Union created a number of new states responsible for ratifying the treaty even as they dealt with fundamental constitutional and security policy issues, the treaty formally entered into force without amendment on November 9, 1992. By mutual agreement actual application on a provisional basis had started earlier, on July 17, 1992.[16]

Conventional Forces in Europe Treaty

The purpose of the Treaty on Conventional Forces in Europe is to establish numerical ceilings for conventional main battle weapons and reduce existing weapons to those limits. CFE applies to Europe from the Atlantic to the Urals, and to all land-based weapons within that area that fall into the regulated categories. The treaty defines five categories of "conventional armaments and equipment:"

- Tanks
- Armored combat vehicles (which includes armored personnel carriers, armored infantry fighting vehicles, and similar vehicles)
- Artillery (both guns and rocket launchers of 100 mm or greater caliber),
- Combat (fixed-wing) aircraft
- Attack helicopters (but not combat support helicopters).[17]

For each category of armaments, equal numerical limits are specified for each alliance area.[18] These limits apply to the geographical area (defined in terms of the territory administered by the members of each alliance when that alliance was established), and not to the alliance—one of which no longer exists—or to particular states.[19] In addition to these treatywide limits, there are separate sets of sublimits. The CFE treaty's geographically complex approach to arms limitations is not, however, the focus of this study. The question is how the CFE treaty handles verification.

As might be expected, this process begins with each party making an initial declaration as to the "maximum levels of holdings of conventional armaments and equipment" limited by the treaty.[20] The treaty actually provides that each state make one declaration of the weapons it possesses at the time the agreement enters into force, and a second declaration after 40 months (during which time each state is to destroy those weapons exceeding its limits) indicating the maximum quantities that state plans to possess in the future. Annual declarations of actual quantities are to be made every December 15th. These notifications are to be made in accordance with procedures established in a Protocol on Notification and Exchange of Information. That Protocol stipulates additional detail for the information to be reported, that each party is to provide all required information to all other states parties, and provides formats for these reports. In addition, parties are permitted certain conditional exceptions to some treaty requirements, but invoking one of these exceptions requires notification directly to each other party.

Article XIV establishes general measures "for the purpose of ensuring compliance with the provisions of this Treaty." To this end, "each State Party shall have the right to conduct, and the obligation to accept, within the area of application, inspections in accordance with the provisions of the Protocol on Inspection."[21] Inspections are to serve three purposes: (a) to verify compliance with the numerical limits; (b) to monitor the reduction process in which excess armaments and equipment is destroyed; and (c) to monitor the process of certifying certain equipment which is modified so as no longer to qualify in controlled categories (rather than being destroyed).[22]

While inspections are to be performed by one party of another, a state party may not inspect another in the same group "in order to elude the objectives of the inspection regime."[23] Thus, one ally may not inspect another for the purpose of filling the second's annual quota of inspections. (As only NATO still exists, the practical significance of this provision is reduced somewhat. Eastern European parties now inspect each other for legitimate security reasons). The number of inspections each party has the right to conduct and is obligated to accept during "each specified time period" is to be established according to provisions of the Protocol on Inspections.[24]

Finally, each party has the right to conduct, and each party within Europe[25] has the obligation to accept, an agreed number of aerial inspections.[26] When the Treaty was negotiated, the intention was to develop an aerial inspection regime for this purpose and put it into effect during the 40-month notification period following the Treaty's entry into force. However, no state subsequently proposed a specifically CFE aerial inspection regime. Whether by design or default, there is general agreement among the parties that the Open Skies Treaty will suffice for this purpose.[27]

In addition to these overt inspections, and also "for the purpose of ensuring verification of compliance with the provisions of this Treaty," a party has the right to use "national or multinational technical means of verification at its disposal in a manner consistent with generally recognised principles of international law."[28] The inspected party may not interfere with national or multinational technical means of verification of another party so long as this condition ("consistent with . . . international law") is met.[29] This includes using concealment measures that "impede verification of compliance."[30] While use of national technical means (NTM) has been a verification feature of bilateral (U.S.-USSR) arms control agreements, this may be the first multilateral treaty explicitly relying (albeit not exclusively) on NTM, much less contemplating multinational use of "technical means"—a term indicating observation or measurement from a distance, as opposed to on-site.

While verification is conducted by each party (although several may conduct an inspection together), some questions are best handled by the parties acting as a group. The treaty establishes a Joint Consultative Group (JCG), consisting of the states parties, which is empowered to:

> (A) address questions relating to compliance with or possible circumvention of the provisions of this Treaty;
> (B) seek to resolve ambiguities and differences of interpretation that may become apparent in the way this Treaty is implemented;
> (E) resolve technical questions in order to seek common practices among the States Parties in the way this Treaty is implemented;
> (G) consider and work out appropriate measures to ensure that information obtained through exchanges of information among the States Parties or as a result of inspections pursuant to this Treaty is used solely for the purposes of this Treaty;
> (I) consider matters of dispute arising out of the implementation of this Treaty.[31]

To date this body has considered questions of treaty interpretation involving inspection site boundary definition and "passive" inspection quotas (that is, how many inspections the party is obligated to accept). It has not been called upon to handle any controversial case involving evidence that a state party is violating its limits on treaty limited equipment. Given the absence of any sanctions authority, and the requirement that the JCG "take decisions or make recommendations by consensus,"[32] it is likely that any confrontation regarding evidence of limit violations would rapidly escalate to the U.N. Security Council.

Protocol on Inspection

As with other agreements we have considered, the CFE addresses verification issues in greater detail in a subsidiary document, in this case one of the protocols to the treaty itself. The Protocol on Inspection establishes "procedures and other provisions governing the conduct of inspections" as provided for in Article XIV of the treaty.

The conceptual framework for CFE verification is very different than that of the nonproliferaton agreements discussed in previous chapters. The IAEA and NPT focus on quantities of nuclear material, with a secondary focus on activities. The CWC verification system and the compliance system emerging for the BWC (which cannot be called a verification system yet) focus on activities and facilities rather than quantities of controlled materials. The CFE treaty, however, stipulates obligations in terms of quantities of specific weapons (numbers of tanks or aircraft, for example) as a practical measure for the real focus, combat capability, and it is at this point several complicating factors intrude. One is that combat capability rests not only on the number of tanks or aircraft, but also on the existence of organizational structure to deploy that equipment in combat. For this reason there is a secondary focus throughout the CFE treaty on military components and organizational structure. This dual level of focus (on items of equipment, and then on organizational capability to deploy undeclared equipment) is reflected in the inspection and verification process.

When the negotiations turned to questions of inspection quotas and what constitutes a site for a single inspection, differing military practice led to different perspectives. Some states station an entire division (or more) at a single large site, while other states station individual regiments (or even smaller units) at a site. The Soviet Union proposed the concept of an "object of verification" to limit access during a single inspection of a Soviet installation to some entity comparable to what could be inspected in several other states. This led to a compromise in which the treaty distinguishes between controlled items (such as tanks and fighter aircraft) and "objects of verification."[33] This is reflected in some unusual definitions:

Inspection Site: an area, location, or facility where an inspection is carried out, but is distinct from both an "object of verification" and "conventional armaments and equipment."

Object of Verification: not an individual item of conventional armament or equipment, but rather:

—a military formation or unit (primarily but not exclusively at the brigade/wing level) which holds relevant conventional armaments or equipment;

—a designated permanent storage site for subject conventional armaments and equipment; or

—a reduction site for conventional armaments and equipment (a place where conventional armaments and equipment in excess of the state party's allotment is being destroyed or converted to other uses.[34]

The inspecting party normally has a right of access to such "objects of verification." Access to other "inspection sites" is contingent.

The concept of a "declared site" in the CFE differs in an essential respect from that used in IAEA/NPT and CWC inspections. In those cases, the inspected state defines (with some limitations identified in the foregoing chapters) the site to be inspected. In the CFE treaty, a "declared site" is to contain one or more "objects of verification" and "shall consist of all territory within its man-made or natural outer boundary or boundaries as well as associated territory" normally utilized by the "object of verification" (military unit).[35] During a single inspection the inspectors "shall be permitted access, entry and unobstructed inspection . . . within the entire territory of the declared site" except for areas "belonging exclusively to another object of verification."[36] Unlike the NPT/IAEA and CWC cases, associated use of an area compels a right of inspector access.

The Protocol identifies several different types of inspections:

- *Declared Site Inspections* are routine inspections of the area where an *object of verification* is located; a state cannot refuse to permit a declared site inspection (except, presumably, when the requested inspection would require that the inspected party accept more than its annual quota of inspections).[37]
- *Challenge Inspections within Specified Areas* are those where the inspecting party defines a geographic area is to be inspected for evidence of armaments or equipment in excess of the state's limit or located in a prohibited area (because it exceeds the zone quota). A request for a challenge inspection can be refused, in which case the "inspected State Party" (this is the terminology of the agreement, notwithstanding the absence of an inspection) "shall provide all reasonable assurances that the specified area does not contain conventional armaments and equipment limited by the Treaty." However, "if such armaments and equipment are present and assigned to organisations . . . to perform internal security functions in the area . . . the inspected State Party shall allow visual confirmation of their presence."[38] It is unclear why a challenge inspection is not obligatory when requested. As there is no independent entity to judge the validity of the request and perform the inspection, as with the CWC, there is no independent check on misuse of the challenge inspection power. On the other hand, when all challenge inspections are voluntary, one must be concerned lest the inspection power becomes a means solely of confirming compliance, without the power to detect noncompliance—a verification Trojan Horse.
- *Inspection of Certification.* A state possessing more combat attack helicopters or fixed-wing aircraft than permitted may convert these to combat support, trainer, or civil internal security applications, and then certify each aircraft as having been converted. The certification process is subject to inspection by any other party.[39]
- *Inspection of Reduction* is an inspection of a site where "the process of reduction" is taking place, that is, where subject conventional armaments and equipment are being destroyed or

otherwise rendered no longer weapons-capable and subject to the treaty limits.

A party cannot refuse to permit inspections of certification or reduction.[40] The treaty stipulates that inspections of certification or reduction may be performed by a team of inspectors from several parties.[41] It is silent on the question of multiparty inspection teams for declared site and specified area inspections, but both have occurred in practice (including even teams including inspectors from both groups of states!).[42] However, all inspection teams are led by a single-state party.

The rights and obligations with respect to inspectors are also defined in the Protocol. One unusual feature is that an inspector must be a national of a party, but need not be a national of the state for which he is conducting inspections.[43] A party may refuse to accept an individual as an inspector for the proposing party (likewise it may refuse to accept an individual as a transport crew member).[44] Parties are obligated to provide visas for inspectors and transport crew members and diplomatic clearance numbers for their aircraft or vehicles. The inspected party has the right to specify entry/exit points, and the inspection team must use those points, although the inspecting state can request to use another point.

The CFE treaty contains some very unusual provisions concerning languages, most of which are not relevant for our purposes.[45] Inspectors are guaranteed the right to communicate among themselves in a language that may not be understood by their escorts[46] and to communicate with their embassy in the inspected party, but "using appropriate telecommunications means provided by the inspected State Party."[47] Hence secure communications between the inspectors and their government, while not prohibited, are unlikely as the inspecting state controls the equipment.

The treaty contains an interesting balance in protecting the rights of the inspected state (not to be subjected to unreasonable numbers of inspections) and the interest in insuring that advance notice does not permit the inspected state time to conceal violations. When one party notifies another of its intent to inspect, it must include certain information about who, how, when, why, and such. The inspected party "shall immediately upon its receipt send copies of such notification to all other States Parties."[48] As there are a number of limitations on the obligation to accept additional inspections, either of the same site, or at other places within the country, this puts all other parties on notice as to whether they may reasonably expect a positive response should they request an inspection of the same party.

While it is important to inform all states of inspections soon to be conducted, it is also necessary for effective inspections that the inspected state not receive advance of information on the specific site, because in the conventional forces case this would permit removal of evidence that limits were being exceeded. For this reason the inspecting state does not identify the site to be inspected until its team arrives at the point of entry, and the object of verification is identified only when the team arrives onsite.[49] The inspected state is required to transport the inspection team to the site immediately (normally within 15 hours) after identifying that site.[50] In the conventional forces case, convincing evidence of a violation can be removed from a site very quickly, hence the importance of short notice to the inspected state. The need to maintain genuine surprise to ensure effectiveness is one reason a single international inspection entity was never seriously considered for CFE.[51]

When an inspection team arrives in-country, the inspected state party has the right to examine equipment and supplies brought by the inspectors. If the escort team determines that "an item of equipment or supplies . . . is capable of performing functions inconsistent with the inspection requirements of this Protocol," the escort team may prohibit its use and impound it for the duration of the inspection.[52]

The restriction on use of information obtained in the course of inspections is quite unusual: "No information obtained during inspections shall be publicly disclosed without the express consent of the *inspecting* State Party." (emphasis added)[53] Thus, the inspected state may not engage in a public relations campaign intended to assert that the inspection produced a "clean bill of health." This provision may also relate to the rights of the inspected party to review the inspection team's report.

The CFE treaty emphasizes resolving ambiguities at the time and place they are identified rather than permitting the inspectors to go home and consult with their own government and perhaps thereby politicize the issue. "States Parties shall, whenever possible, resolve during an inspection any ambiguities that arise regarding factual information."[54] The inspection team's escort is to respond promptly to requests for clarification. Inspectors may also document an ambiguity with photographs "using a camera capable of producing instantly developed photographic prints," provided that the escort agrees.[55] Ambiguities not resolved are to be documented, with explanations provided and photographs taken, in the inspection report.

The inspection report provisions of the Protocol are most unusual. "In order to complete an inspection .. and before leaving the inspection site: (A) the inspection team shall provide the escort team with a written report, and (B) the escort team shall have the right to include its written comments in the inspection report and shall countersign the report within one hour after having received" it.[56] Inspection reports are to be "factual and standardised."[57] This "no surprises" feature means that the inspected state is immediately informed of any anomalies (and by countersigning the inspection report certifies this knowledge), and has the opportunity to provide an explanation (which may resolve the anomaly or dig the hole deeper).

Analysis

The Conventional Forces in Europe treaty appears to differ in many fundamental respects from the previous treaties (and regimes) considered. In fact, here there is no "regime" or effort to build one, in terms of the definition in chapter 1. The CSCE agreements provide a larger political framework, but not in the same sense the term "regime" was used in preceding chapters. The CFE treaty only *limits*, not *prohibits*, possession (and possibly the use) of specified weapons systems. Nor does CFE in any way address the production of those weapons systems (although it does address in detail the destruction of excess treaty-limited weapons), and for this reason does not involve any verification of civil activities that might be used for covert weapons production.

The CFE treaty is regional in scope. This difference with the nonproliferation treaty regimes may be more apparent than real. Each of those treaties is intended to apply to all states, to cover the world. CFE is intended to apply only to Europe and to address a security issue that can, to a significant degree, be limited in this geographic sense. Perhaps the more significant difference is that CFE was negotiated among all the relevant states (with the full expectation that when concluded all would become parties). CFE is truly universal within the context of its limited security "universe." And that universe, at least when negotiations began, could be divided into two alliances. This limited size and internal structure must have strongly influenced the negotiators' perceptions of the verification problem.

From the perspective of this study, the Conventional Forces in Europe Treaty utilizes an unusual verification approach to address an unusual verification requirement. The scope involves only military, not civilian, activities and is largely focused on possession rather than use or production. There is a secondary focus on military structure and organization. Quantities of tanks or fighter aircraft are relatively easily measurable (just as quantities of nuclear materials are relatively easy to measure). However, inspecting military organizations would seem to be much more difficult, just as identifying the true character of certain nuclear or biological or chemical activities is exceedingly difficult. The technical characteristics of the verification problem can be satisfactorily handled by the essentially bilateral verification approach adopted. Further, because the universe of relevant states is reasonably small and culturally consistent, a bilateral approach is practicable.

Of equal importance is the existence of previous CSCE agreements on much the same security concerns that also involved each state party independently inspecting other states party as it saw fit. However, unlike previous CSCE agreements, the CFE agreement is, or at least started out to be, an explicitly bilateral NATO-WTO agreement, and a treaty rather than a nonbinding political agreement. Perhaps for these reasons the United States proposed, early in the negotiations, that for this treaty verification would be performed by each of the treaty organizations, acting on behalf of its members (that is, inspections would be performed by multinational groups of inspectors acting on behalf of the North Atlantic and Warsaw Treaty Organizations, rather than on behalf of their individual states). Such an approach was in essence to "scale up" to the alliance level the same approach long advocated by the United States for bilateral (U.S.-USSR) arms control agreements.

But even as the United States was proposing to adopt an historically tested approach (especially in the Intermediate Nuclear Forces agreement, which had been concluded only months before the CFE negotiations began), the political assumptions underpinning that approach were being invalidated. As the Warsaw Pact dissolved, and some of its former members decided that they were as interested in verifying each other (or at least the Soviet Union) as they were in verifying Western compliance, the appropriate option set changed. It appears that the option of a single entity to perform all inspections was never considered,[58] and attention simply shifted to an approach in which each party has the right to inspect any other. The continued existence of NATO was largely ignored.

It is somewhat difficult to place the CFE case in the "historical contingency/technical substantive" dichotomy. Clearly the verification approach adopted for this treaty reflects historical contingency, but it also reflects the substantive political context and the specific technical character of the undertakings to be verified. Negotiating and ratifying a treaty are fundamentally political acts, as is including a verification mechanism in the treaty. Ultimately the CFE Treaty demonstrates how much political context can determine the technical substance—who is concerned about whose compliance. In the beginning, it was each alliance concerned about the compliance of the opposing alliance. By the time CFE was concluded, the concerns were much different: the compliance of one party was the primary concern of all others, and that party was most concerned about the compliance of an opposing alliance (although Moscow is also concerned, especially in the longer term, about the compliance of states on its own borders). An interesting hypothetical question is what the CFE verification procedures would look like had the earlier negotiations failed, and the treaty was being negotiated today.

Notes

1. U.S. Arms Control and Disarmament Agency ["ACDA"], *Arms Control and Disarmament Agreements* (Washington, DC: GPO, 1990), 319.
2. Ibid., 320.
3. Ibid., 321.

4. Ibid., 321.

5. Nicholas A. Sims, "The Second Review Conference on the Biological Weapons Convention," in *Preventing a Biological Arms Race*, ed. Susan Wright (Cambridge: The MIT Press, 1990), 269.

6. ACDA, 321-322.

7. Thomas Graham, Jr., "The CFE Story: Tales From the Negotiating Table," *Arms Control Today* 21, no. 1 (January/February 1991): 9.

8. Ibid., 9.

9. Interview with Arms Control and Disarmament Agency official, November 5, 1993.

10. Ibid.

11. Ibid.

12. Interview with Department of Defense official, February 23, 1994.

13. The day the treaty went to printing, the text had to be changed to reflect the fact that Bulgaria ceased being a "People's Republic" and became a "Republic." Graham, 9.

14. Interview with DOD official.

15. See, for example, Lee Feinstein, "Russia Asks CFE Partners to Allow Increase in Arms on Southern Border" Arms Control Today 23, no. 9 (November 1993): 25.

16. U.S. Arms Control and Disarmament Agency, *Fact Sheet*, "A Chronology of Arms Control and Related Treaties and Agreements including Confidence- and Security-Building Measures, and Measures Related to Non-Proliferation, Transparency and Defense Conversion," December 20, 1993.

17. Treaty on Conventional Armed Forces in Europe (CFE), Paris, November 5, 1990 (reprinted by ACDA), Article II.

18. CFE, Article IV. The terminology "alliance area" is not found in the treaty, but facilitates understanding it.

19. Ibid., II.

20. Ibid., VII.

21. Ibid., XIV/1.

22. Ibid., XIV/2.

23. Ibid., XIV/3.

24. Ibid., XIV/5.

25. That is, Canada and the United States are not subject to any inspection, only their forces in Europe may be inspected. Russia (originally the Soviet Union) east of the Urals, and parts of Turkey and Kazakhstan, are likewise excluded.

26. CFE, Article XIV/6.

27. Written comments provided by Arms Control and Disarmament Agency official, July 1994.

28. CFE, Article XV/1.
29. Ibid., XV/2.
30. Ibid., XV/3.
31. Ibid., XVI/2.
32. Ibid., XVI/4.
33. Comments by ACDA official.
34. CFE, *Protocol on Inspection*, Section I.
35. Ibid., I/1 (N).
36. Ibid., VI/23.
37. Ibid., VII/1.
38. Ibid., VIII/9/(A).
39. Ibid., IX.
40. Ibid., X/1.
41. Ibid., X/1.
42. Comments by ACDA official.
43. CFE, *Protocol on Inspection,* Section III/2.
44. Ibid., III/4 & 7.
45. Another unusual feature of the CFE Treaty is that each State party is to inform all other parties, within 90 days of signing the Treaty, which of the official CSCE language(s) are to be used by teams conducting inspections of its conventional armed forces. (Article III/12) However, other provisions in the Protocol permit the inspecting State party to stipulate the language to be used in the inspection. (Articles XII/6; IV/2(G) & /3(F)) Presumably the inspected State party lists the languages it will accept, and then the inspecting State party selects a language from that list.
46. CFE, *Protocol on Inspection*, Section VII/9.
47. Ibid., VII/11.
48. Ibid., IV/5.
49. Ibid., VII/7 & 9.
50. Ibid., VII/8.
51. Comments by ACDA official.
52. Ibid., V/5 & 6.
53. Ibid., VI/10.
54. CFE, *Protocol on Inspection, Section* VII/38.
55. Ibid., VII/38 & 35.
56. Ibid., XII/1.
57. Ibid., XII/2.
58. Interviews with Department of State official, December 7, 1993; interview with Department of Defense official, February 23, 1994.

6. ❖ Looking for Patterns

The principal question posed for this study was, to what degree are the verification measures incorporated in multilateral arms control and disarmament (nonproliferation) treaties determined by the randomness of political events within and among governments and to what degree are they determined by the substantive nature of the issue? Previous chapters examine what verification is and why states commonly seek to include verification procedures in multilateral agreements, and how this process worked and what results it produced in four particular cases. This chapter compares how the various causes and processes worked in the development of these four verification systems and begin to derive some more generally relevant findings about the process.

As discussed in chapter 1, two separate factors are discernible in the substantive nature of the issue and its influence on the development of a treaty . One is the character of the policy problem being confronted in negotiating the treaty. This refers to the security problem leading individual states to the negotiating table, as well as the relationships among those states participating in the negotiations. For example, the NPT negotiators faced a policy problem—preventing the spread of nuclear weapons beyond the five declared nuclear weapon states—very different from that confronting those negotiating the CFE treaty—limiting conventional military forces in central Europe. There were also differences in the policy context. In the NPT case, the five nuclear weapons states were unwilling to give up their own weapons (with a number of nonnuclear weapons states supporting, however quietly, this position), but sharing with the great majority of other negotiating states the objective of preventing the further spread of nuclear weapons. In the CFE case, the negotiating parties were at the beginning of the negotiations organized into two military alliances, each centered on one of the two super powers. One super power was only partially located in Europe and the other only stationed troops in Europe. By the end of the negotiations the policy context was, as discussed previously, fundamentally different.

The second factor involves the technical characteristics of the situation. Do the technical characteristics of the policy problem facilitate verification? Can the security objectives be defined in terms of behaviors for which there are technically achievable verification methods? The concept of verification implies a distinction between two sets of activities:[1] one set prohibited by the treaty, and another set of closely related but permitted activities. The verification problem is to distinguish between the two. This requires:

- Access to places where the prohibited activities might be conducted (or possibly other locations) to collect information bearing on whether or not prohibited activities are in fact being conducted
- The capability to distinguish between prohibited and permitted activities once access (information) is obtained.

Normally, access entails sending people to a particular site, but need not require onsite inspection, nor does it necessarily entail access to places where the permitted activities are conducted, although such access is commonly part of verification systems.

A governmental activity, or a new, highly regulated industry heavily dependent on government support, presents an easier verification "target" than does a highly developed, widespread, and relatively unregulated industry. Another aspect of the "technical characteristics" factor is reflected in the idea that "the easier it is to measure, the easier it is to verify." Even when the policy problem is straightforward, verifying compliance with a treaty addressing that problem may not be easy, or even possible.[2] The NPT provides one example: verifying compliance with the undertaking not to develop or acquire nuclear weapons proved very difficult; verifying that civil nuclear programs are not being used directly to support weapons programs proved relatively easy, but diverted attention from the real issue.

In this chapter we will seek to pick apart the various threads of historical contingency, policy problem, and technical practicality to determine how each contributed to create the fabric of a particular treaty. From this we will seek to identify useful verities to guide future treaty negotiation efforts. Logically one would expect the policy problem to be the single most important factor in shaping a regime and its verification system. Technical aspects of the problem should also be important, but perhaps more in terms of constraints and opportunities than as fundamental forces shaping the verification regime. Historical contingency, the random events ranging from exceptionally powerful (or weak) personalities in key roles to external events like the outbreak of war or the sudden collapse of an empire, may play a variety of roles. Some historical contingencies (such as the personalities of key negotiators) will channel the flow of events (as technical factors do), while other historical contingencies may play an opposite role, seeming to intervene suddenly and shape events in unexpected ways.

The Policy Problem as the Central Shaping Influence

Do the four cases just studied support the conclusion that the single most important influence shaping the development of a verification system is the type of policy problem being addressed? A look at the development of these four systems reveals that in the first three cases the original objective was, at least putatively, global disarmament with respect to a particular class of weapons.

The Chemical Weapons Convention provides the purest example of the political problem shaping the verification system as a means to that end. The verification system developed in the CWC supports the contention that the states negotiating this convention were "convinced that the complete and effective prohibition of the development, production, acquisition, stockpiling, retention, transfer, and use of chemical weapons, and their destruction, represent a necessary step towards the achievement"[3] of objectives identified in the preceding clauses of the preamble. The CWC contains measures to address each of the acts to be foresworn. Except for "transfer," these measures appear to be as intrusive and effective as any set of verification measures ever incorporated in a disarmament treaty aspiring to universal adherence. Given the technical complexities of the problem, it is fortunate there was such unity among the negotiators on the character of the policy problem. For all major participants in the negotiations, the putative agenda was the real agenda.

According to the preamble and first articles of the BWC, global disarmament was also the policy objective pursued in negotiating the Biological Weapons Convention. But was this the case? The BWC does not contain any verification system (although it does have rudimentary compliance measures). If our model for why states include verification in multilateral treaties is correct, there should be more verification in the BWC than just simple procedures for one state consulting with another on "concerns." Is this a product of the policy problem being incorrectly identified? Were all states in the negotiation working the same policy problem, or were some states negotiating for reasons other than to produce an effective agreement? Should we search for the reason in technical practicality, or even historical contingency?

One hypothesis is to take the BWC preamble at face value. The policy problem was the potential proliferation of biological weapons. From this perspective, the explanation for why the treaty lacks a verification system is provided by President Nixon's explanation for the U.S. unilateral decision to eliminate its offensive biological weapons program: militarily useful biological weapons are not really achievable. If this is the case, there is no real need to create an elaborate international system to verify compliance because the violator would be wasting precious resources in a fruitless exercise.[4]

A second hypothesis sees the political context of the BWC negotiations differently, as primarily bilateral with a multilateral subtext. From this perspective the United States and the Soviet Union were engaged in a larger process of achieving agreement on a range of arms control agreements. A temporary hiatus in this process had developed, and the two superpowers pursued a BWC as a means to maintain momentum. While other states desired a BWC with real verification procedures, the superpowers needed an agreement. Dealing with the exceptionally difficult technical issues involved in effective verification of biological disarmament would delay that agreement, perhaps for many years. The real policy problem for the superpowers was momentum, not biological weapons. When the super powers agreed, the rest of the international community could only follow.

A third hypothesis, like the first, accepts the proposition that the real policy problem was biological weapons. However, it assumes that the policy problem for the United States was not really the general proliferation of biological weapons, but more specifically a Soviet program to develop effective offensive biological weapons. Most states participating in the negotiations were concerned generally with proliferation of biological weapons, and for the reasons described in chapter 1 these states sought to include some verification measures in the treaty. The United States recognized that the Soviet Union would not accept the intrusive measures necessary to obtain satisfactory verification, but at the same time, the United States wanted a legal commitment that, if violated, would clearly place the Soviet Union outside accepted norms of international behavior.[5] Under this hypothesis the United States hijacked the convention by obtaining early agreement on a text that contained little in the way of compliance measures, not withstanding the larger interests of most states in the negotiation. Reciprocally, the Soviet Union wanted a BWC as a cover for its secret biological weapons program and to disarm its adversaries; the Soviet Union's long-standing opposition to all onsite inspection simply provided a credible basis for opposing verification in this case.[6]

How does one select among these alternative hypotheses? While an extensive and detailed historical analysis could shed considerable light on the degree to which each hypothesis is correct, the important point is that each of these hypotheses reflects one facet of the actual case. There was no single policy problem that all participants in the BWC negotiating process were working equally. Many states pressed hard for verification measures, and Britain proposed a text with relatively strong measures. In the end, all participants agreed on a text with no verification measures as such, and only very weak measures to address questions of compliance on a bilateral basis. We will explore the reasons for this further when we turn to the role of technical practicality.

As is clear from chapter 2, the nuclear case reflects a long evolution, agreement building on agreement (or lack thereof). In 1946, the United States and most other participants in the United Nations Atomic Energy Commission were seeking to eliminate nuclear weapons. Not only did this initial effort fail, it failed so dramatically as to force a substantial redefinition of the policy problem, from global nuclear disarmament to nonproliferation—that is, restricting possession of nuclear weapons to those states already possessing them.

The Baruch Plan failed because of a fundamental reluctance to cede sovereignty to an international organization. One cannot know what would have happened had the Soviet Union not rejected it, but it is unlikely that negotiations, and the process of implementing the final agreement, would have produced a regime exercising the extensive supranational powers proposed by the United States. There is no reason to believe that the Soviet Union was the only member of the new United Nations that did not agree with the U.S. proposal, and it is uncertain whether the United States would have ratified a treaty incorporating the Baruch Plan.[7] Experience with the IAEA demonstrates how strongly states will resist actual erosion of sovereignty, even after agreeing in principle.

While Atoms for Peace reversed the priority previously given to prohibiting nuclear weapons and developing civil uses of nuclear energy, in some respects this was (at first) a tactical move. President Eisenhower knew he could not roll the clock back; the United States, the Soviet Union, and Britain would continue to have nuclear weapons. With this caveat, his first conception of the international organization was remarkably similar to the Baruch Plan. All cooperation in civil nuclear uses would go through the IAEA, which would retain some control over items transmitted, plus apply safeguards to ensure that materials and technology were not diverted to military applications.This scheme was drastically modified on the way to realization.

When the United Nations returned to the issue of nuclear proliferation in the mid-1960s, the context was much changed. Countries such as India, Argentina, Germany, and Japan now had extensive and partly (in some cases largely) indigenous nuclear programs. As a consequence, the policy context was different. Just as in the mid-1950s when the IAEA was established, the objective (for most states) was not to eliminate nuclear weapons, but to stop further proliferation (a few participants in the negotiation still sought disarmament, or at least sought what would be for their security concerns a level playing field).[8] The focus was on creating an effective fire-break between civil nuclear programs, which were to be fostered, and nuclear weapons programs, which were to be prohibited. The focus on fostering civil nuclear activities diverted attention from the problem of verifying the absence of solely military programs. In essence, many negotiators confused confirming that permitted activities are not being misused with actually confirming that prohibited activities are not being conducted. As with the BWC, different states participating in the negotiation defined the policy problem differently, because they had different reasons for being there at all. In the end each focus was accommodated and none entirely achieved.

The Conventional Forces in Europe agreement, like the CWC, addresses a relatively straight-forward policy problem widely, if not unanimously, shared among the negotiating states. The objective was to reduce conventional forces in central Europe and to do so in a fashion that would genuinely increase confidence and security for all states in the region. Many years had been spent attempting to achieve this objective by negotiating reductions in military personnel, an approach that had consistently foundered on definitional and verification difficulties. Finally the idea emerged to focus instead on limiting quantities of main battle armaments.

Approaching the issue in terms of main battle armaments was not without its own problems. The intent was to limit offensive weapons without unnecessarily limiting defensive weapons, thus the early focus was on tanks, armored vehicles, and artillery. However, attack helicopters are not only antitank (defensive) weapons, they are also effective substitutes for tanks. Attack aircraft have similar multiple roles. The weapons and equipment to be covered by the agreement and appropriate tradeoffs among these items were a primary focus of the negotiations.

While the numbers of each item that each state is permitted to possess are set precisely, it was clear to the negotiators that no practical verification system could distinguish between, for example, 500 tanks and 530 tanks. However, there is no military significance to minor deviations above permitted levels, and verification could be made sensitive enough (by incorporating national technical means as an element) to identify violations of military significance. Like the CWC, the CFE verification system is clearly consistent with the policy objective of the treaty.

Thus we can say that the policy problem plays an important role in shaping the political and legal undertakings in the treaty, that is, in determining what might be verified. The definition of the policy problem is also important to determining whether there is a verification system in the treaty, and perhaps in influencing how rigorous it is. There are still, however, a great many differences among our four cases remaining to be explained.

Historical Contingency

Contingent events clearly play a role in the development of all international agreements. The issue for this study is whether verification systems are primarily shaped by substantive factors, or does historical contingency play such a large role as to render the substantive factors potentially inconsequential?

The Chemical Weapons Convention demonstrated the importance of random (at least in our context) historical events. In contrast with the BWC negotiations, the Western powers entered the CWC determined that any treaty must include effective verification measures. Absent measures to ensure that the USSR would not flout the CWC as it had the BWC, there would be no agreement. For the first several years there was every reason to believe that this meant no agreement, but political events in the Soviet Union intervened. The selection of a new premier, Mikhail Gorbachev, who reversed decades of Soviet policy and agreed to, even advocated, intrusive onsite inspections as part of a rigorous verification system, permitted these negotiations to progress from political theater to genuine substance and ultimately to the CWC. Historical contingency made the CWC achievable, but it is not accurate to say that this historical contingency shaped the verification system.

Political events within a government did, however, play an important role in the development of the CWC verification system. Efforts by one faction in the U.S. Government to make the negotiations fail and place the blame on the Soviet Union led to the U.S. proposal in 1984 for challenge inspections "anywhere, any time." This proposal went beyond what even the United States was willing to accept. It shifted the subsequent course of the negotiations, but not as expected by its advocates. Instead of working (conceptually) from sovereignty towards verification, the negotiators started from highly intrusive verification procedures and worked to make those acceptable. The result was the most intrusive multilateral verification system ever agreed to by modern states.

It may appear harder to evaluate the role of historical contingency in the BWC. The United States approached negotiation of this agreement with several policy objectives that were to some degree in competition. The Soviet Union appears to have approached this negotiation with at least two objectives: maintaining momentum in the arms control process, and getting an agreement that would hide rather than detect the Soviet Union's offensive biological weapons program.

From the perspective of most states participating in the BWC negotiations, historical contingency was central. They were seeking an agreement that would eliminate biological weapons and include some reasonable verification measures. The British tabled a text including compliance verification, and several other states strongly supported that text, but the two super powers simply were able to win at the negotiating table. Whatever the story with respect to arms control momentum, neither super power entirely achieved its objective with respect to treaty compliance. The USSR did not find the BWC an effective cover, and the United States found the treaty a double-edged sword as a diplomatic tool. As U.S. accusations of Soviet cheating continued to resonate but escaped any meaningful resolution, Britain, Sweden, and the other states seeking a more effective compliance system were gradually able to build more and more compliance machinery into the BWC regime. Whether this will lead to a real verification system remains to be seen.

The IAEA and NPT story is full of historical contingencies which shaped the verification regime. The single most important of these was Atoms for Peace; President Eisenhower sensed that the situation called for radical approaches, and he developed his own.[9] But Atoms for Peace proved so popular that the United States began bilateral cooperation, supplying research reactors, extensive training, and nuclear fuel directly to other countries at least 2 years before the new IAEA was established. In negotiating the statute for this new organization, many states expressed strong opposition to granting it the extensive inspection and control rights originally envisioned by

President Eisenhower. By providing the intended fruits of cooperation on different terms, and before the originally intended arrangements had been established, the United States undercut its own plan. Enthusiasm for parts of the plan, even in a formerly very dour Congress, had the effect of changing the policy problem.[10]

Historical contingency also shaped the NPT safeguards system. Although this system is designed to verify that no NPT party is violating its treaty commitments, the system was created at a time when most of the nonnuclear weapon states with significant civil nuclear programs were allied with the United States and not considered likely to seek nuclear weapons in any case. These states were more concerned with "guarding against the risk that the distinction between weapon and non-weapon states inherent in the NPT would extend to peaceful nuclear activity" than they were with the (at that time) remote possibility for a real nuclear-armed renegade.[11] Proliferation by Iraq and North Korea were some 25 years in the future; even South Africa, Israel, and India (and Sweden) were fully within the fold (or were at least widely perceived to be so) in the mid-1960s.

Historical contingency intervened in 1990 to correct this original, overly soft approach to verification. Saddam Hussein invaded Kuwait, then repeatedly misread the unity and determination of those aligned against him, resulting in Iraq's defeat in the Gulf War. Only then did the magnitude of Iraq's NPT violations come to light. The discovery of multiple undeclared uranium enrichment programs, some very close to producing significant quantities of highly enriched uranium, galvanized an international consensus that IAEA safeguards had failed. It also forced the recognition that this failure was due primarily to the constraints imposed on safeguards by the NPT parties themselves. This realization led to a new consensus in the IAEA's Board of Governors and to approval of important reforms. Historical contingency will also determine whether the Board of Governors (and member states) maintain the political will to ensure that these reforms are fully implemented and institutionalized, or whether most Governors revert to the old way of doing business.

The Conventional Forces in Europe negotiations started out as an essentially bilateral process between NATO and the Warsaw Treaty Organization. Verification was originally conceived as being performed by the two alliances, each inspecting the other. The breakup of the WTO during the negotiations fundamentally changed the nature of the verification issue for most parties. The result was a system in which each state may verify compliance of any other. Verification responsibilities shifted from alliances to individual state.

Alternatively, one might argue that nothing really changed at all. Verification was always to be the responsibility of the individual states parties. While inspections might consist of NATO and WTO teams of military officers, neither NATO nor WTO had an independent secretariat similar to IAEA or the United Nations. The political alliances among parties changed during the negotiations, and this fact was recognized, but it did not change the verification process in any important way. Washington and Moscow always intended that judgments on the inspection findings would be made in Washington and Moscow, not Brussels and Warsaw. Historical contingency may have loomed large for the negotiators and changed the political significance of the agreement in important respects, but for the verification system it really meant little.

The CFE case demonstrates how difficult it can be to determine whether historical contingency was a significant factor or not. Certainly Russia is being inspected by states that would never have considered inspecting the Soviet Union when they were alliance partners in the WTO, but it is likely that the verification system itself is really very similar to what might have been developed between the two alliances.

One important element of historical contingency for the CWC and CFE treaties was the role of the U.S. Congress. Experience with previous multilateral treaties (including the NPT and the BWC) and bilateral arms control agreements with the Soviet Union had convinced most members of Congress that rigorous and implementable verification measures were necessary in the CWC and CFE agreements. As the United States was essential to each treaty, all other states participating in the negotiations understood that they must either work with U.S. negotiators on verification measures that would pass congressional scrutiny or forego their own political and security objectives in a treaty.[12]

Historical contingency is clearly a major influence on the shape of the verification system negotiated for a particular multilateral agreement. But it does not seem credible that this is the whole story, and there remain a number of differences among the verification systems examined in the foregoing chapters which are not explainable in these terms.

Technical Practicality

The third factor in our analysis deals with questions of what is possible given the technologies of the times. Specifically, what kinds of verification are practical and what kinds of verification may prove unachievable in practice. It was suggested in the introduction to this chapter that questions of technical feasibility will establish constraints and opportunities but will not shape verification systems to the degree that the previous two factors do. Is this hypothesis supported by the four cases examined?

One central difference among these cases is the nature of the activity to be controlled. In each case the agreement prohibits certain activities but permits closely related activities (in the case of CFE, it permits a particular activity up to a defined threshold, and prohibits the identical activity above that threshold). Differentiating permitted from prohibited activities is the *sine qua non* of verification and will depend on many technical characteristics of the activities in question. One is the ability of the verification authority (whether international or national) to obtain information about permitted activities. This in turn will depend on whether the activity is performed by the government (e.g., conventional arms) or commercial firms (e.g., chemical and biological industries), and if the latter, the degree to which governments already regulate the permitted activities, and the degree to which the activity is nascent versus well developed.

CFE addresses military and state security activities, and so verification entails access only to well-established functions performed by the state. Rights of citizens and commercial sensitivities are not a problem. By contrast, the civil nuclear industry was in its infancy when the IAEA Statute was negotiated and still quite undeveloped when the NPT was negotiated. In addition, there evolved from the beginning a sense that civil nuclear activities should be closely regulated by the state. For these reasons, in the conventional arms and nuclear cases verification of permitted activities should have been relatively easy to negotiate, as the principle of state access already existed. Recalling that even "relatively easy" may equate to difficult, this proved to be the case. How the NPT will ultimately compare with the CWC (and perhaps the BWC) cannot be determined until years of experience in actually implementing the latter treaty(ies) is gained.

The chemical and biological industries, by contrast, are private commercial activities in most countries and were well-developed industries long before efforts to negotiate international controls began. Neither industry has traditionally been subject to the kind of extensive regulation imposed from the beginning on the nuclear industry. In these industries access for the verification authority has proven more difficult to obtain. Nonetheless, CWC offers the most extensive access of the four agreements. While many years of negotiation were needed, a consensus existed among the negotiators that such access was necessary, and acceptable methods to provide that access were found. However, such also appeared to be the case for the NPT in 1972, and success in the CWC cannot yet be claimed.

Distinguishing permitted from prohibited activities arises once the inspection takes place. In the CWC and the NPT, distinguishing between acceptable civilian activities and prohibited weapons-related activities is in fact difficult. Many chemicals with large volume commercial uses are also either potential weapons agents or are the immediate precursors for such agents. Similarly, certain normal operations in civil nuclear programs produce nuclear materials usable in nuclear weapons, and other normal activities are difficult to distinguish from covert production activities. In each of these cases intrusive onsite inspections are necessary at civil facilities to confirm that prohibited activities are not being conducted. Were some form of effective remote surveillance possible, each of these verification regimes might be very different.[13]

Nuclear materials differ from chemicals in that they emit radiation, and measuring that radiation can provide information concerning the specific kind and quantity of nuclear material in a sealed container. In addition, relatively small quantities of certain nuclear materials are all that is needed for one nuclear weapon. For these reasons the IAEA verifies plutonium and enriched uranium to the gram. One reason that IAEA safeguards came to focus so heavily on material accountancy may be that measuring nuclear materials is relatively easy; another is that a material accountancy approach minimizes the degree to which safeguards interfere with or permit insight into the actual operations within a plant.[14] Minimizing the inspector's knowledge of plant operations protects commercially sensitive information; in certain cases it also helps control technologies of great proliferation significance.

A militarily significant quantity of chemical agent is measured in the hundreds or even thousands of tons. This is fortunate, as in a commercial chemical plant it is impractical to account for chemicals used in and produced by the process to the nearest 10 kilograms. This is one reason CWC verification does not focus on materials accountancy. (Another is the IAEA experience with excessive and, some would argue, misplaced focus on accounting.) This technical impracticability may help protect the OPCW from being forced down the same false path taken by the IAEA.

The BWC presents a very different situation from either of the foregoing. Militarily significant quantities of biological agents are relatively small and can be fabricated in a modest facility that may be identical to a permitted civil facility (in fact, a vaccine factory could be used to fabricate offensive warfare agents by simply omitting the last few steps and substituting others, perhaps at a different location). Further, these quantities can be produced in a relatively short time (and are not demonstrably for prohibited rather than permitted purposes until the 11th hour). These factors make verification of a BWC much more difficult and uncertain than is the case for either nuclear or chemical weapons. Until very recently the United States took the position that a verification system for the BWC is more likely to provide a false sense of assurance than an acceptable level of confidence that violations are not taking place,[15] a matter being revisited by the BWC parties.

The technical characteristics of the verification problem are significant, but certainly not to the extent that one might have imagined. The technical characteristics of the problem shape the tactics of verification but not the strategy of verification. That appears to be more determined by broader political factors (such as the degree to which states are willing to delegate responsibility in the particular instance) and the historical situation, such as whether there is a pre-existing international organization or a pre-existing fabric of related agreements relying on bilateral verification.

In sum, the type of policy problem (including whether there is consensus on its definition), the technical characteristics of the verification problem, and random historical events all play important roles in shaping each verification system. But what conclusions might be drawn from this?

Notes

1. Previously the term "activities" was used in a somewhat narrower sense than it is used here. In this case the "activity" may be the production of plutonium or an organic chemical (as it was in previous chapters), but in this discussion "activity may also refer to possessing 1000 main battle tanks, while a distinguishably different "activity" would be to possess 1080 main battle tanks.

2. As was indicated in the discussion of what verification is (chapter 1), "adequate" or "effective" verification is, like beauty and pornography, often in the eye of the beholder.

3. Chemical Weapons Convention preamble.

4. This analysis addresses the issue of verifying compliance with respect to militarily significant quantities of biological and toxin agents. Much smaller quantities could be produced by a state for use in covert/terror operations. Detecting such small scale production would require vastly more intrusive verification methods, and if the possibility of undeclared production laboratories is considered (as it must be) is essentially beyond the capability of any practical verification system.

5. If a state has not undertaken a commitment to forego development or possession of biological weapons, efforts to make discovery of a covert biological weapons program would appear to most other states as simply more evidence of the fundamental political conflict. However, if that state has a formal legal commitment not to develop or possess biological weapons, detection of a covert program signals that the state does not honor its commitments. The terms of the issue are changed from developing a particular kind of weapon to breaking one's commitments. We see the same process with respect to North Korean adherence to the NPT but refusal to accept special inspections by the IAEA.

6. This is based on Arkady Shevchenko's claim (see chapter 2) that the Soviet Union had an offensive biological weapons program at the time the BWC was negotiated and never intended to terminate that program. Rather

it sought military advantage by cheating, so it sought to minimize the risk of getting caught.

7. The mood of the U.S. Congress was reflected in the draconian Atomic Energy Act of 1946, totally eliminating all nuclear cooperation with the United Kingdom, Canada, and France. The cooperation of these three allies had been essential to U.S. development of the atomic bomb. Given this mood, it is unlikely that Congress would have agreed to ratify and implement a treaty based on the Baruch Plan, leaving President Truman in much the situation President Wilson had found himself 25 years earlier.

8. Virtually no state except India argued that the NPT negotiations should have been about global nuclear disarmament instead of prohibiting further proliferation of nuclear weapons. The simple explanation is that a great many, if not most, of the states participating in the negotiations benefitted from the nuclear umbrella of either the United States or the Soviet Union. Their security interest was in preventing potential adversaries from obtaining nuclear weapons, not in eliminating the balance of strategic deterrence between the super-powers. India had a very different security problem, the principal threat to its security came from China, which was already a nuclear weapons state. For India the only satisfactory situations were Chinese nuclear disarmament, sanction for India to obtain nuclear weapons, or some very credible and powerful security guarantees for India. Only the second option appeared to be within India's power to achieve.

9. See Lawrence Scheinman, *The International Atomic Energy Agency and World Nuclear Order* (Washington, DC: Resources for the Future, 1987), 61-62.

10. Once Atoms for Peace was being implemented, most of the international community was more interested in obtaining the formerly prohibited technology than they were concerned about the prospect of the further proliferation of nuclear weapons. Whether this was the case before Eisenhower's speech cannot be demonstrated, as the opportunity was not there. Certainly the United States was more interested in arms control than the expansive distribution of civil nuclear technology, that is, until the political benefits of the latter became so apparent.

11. Lawrence Scheinman, "Lessons From Post-War Iraq for the International Full-Scope Safeguards Regime," *Arms Control Today* 23, no. 3 (April 1993): 3.

12. I am grateful to Leonard S. Spector of the Carnegie Endowment for International Peace for raising this point in his comments; letter to the author dated July 5, 1994.

13. Effective and acceptable (two independent and very rigorous criteria) remote surveillance methods would, if capable of collecting all the necessary information, obviate the need for onsite inspection. Whether such remote surveillance methods would be judged less intrusive and more acceptable on sovereignty (and other) grounds is questionable. Much research and development effort has been devoted to remote sensing techniques for nuclear safeguards, and in recent years remote surveillance has become a useful adjunct to, but not yet a replacement for, on-site inspections. Two major problems entail ensuring that the remote surveillance technology is not being spoofed, and national concerns about protecting commercial information and protecting employee privacy (ironically this last argument is most frequently advanced by states with little domestic record for such concerns).

14. Many nuclear safeguards and nuclear industry experts might take umbrage at this characterization of accounting for nuclear materials as "easy," but the fact remains that it is technically practical in a way that accounting for chemicals in similar industrial applications simply is not. It is also the case that nuclear materials accountancy has grown steadily with the evolution of the nuclear industry. John McPhee has shown (in *The Curve of Binding Energy*) how primitive nuclear materials accounting practices were during the 1950s and 1960s, but even these practices were vastly more precise than what is routine (or affordable) in the chemical industry today.

15. Experience with the NPT in Iraq, plus extensive reporting of Soviet cheating on the BWC, provide credence to the argument that a false sense of assurance, once shown to be false, may be as (or more) damaging to a treaty than recognition that it contains no really effective compliance verification measures. Note in particular the rapid erosion in public confidence in the IAEA and the NPT following Israel's bombing of the Osirak reactor and its accusations that Iraq had a nuclear weapons program undetected by IAEA safeguards.

7. ❖ Conclusions

The foregoing chapters examine what verification is and why states would bother with so difficult and politically sensitive an issue when negotiating agreements on arms control and disarmament issues. Now it is necessary to confront the question of whether there are any meaningful conclusions to be drawn from this exercise. Are the patterns discerned in the history of these treaties meaningful for understanding how other treaties have evolved or will evolve? Are there lessons here which might benefit future negotiators? This final chapter seeks to provide some answers, albeit partial ones, to these questions. There are in fact several interesting and potentially important conclusions to be drawn.

Verification of multilateral treaty obligations contains its own intrinsic structure and logic, independent of the obligations undertaken by the parties and the political context in which those undertakings are negotiated and made. The many significant similarities in the verification processes for the CFE Treaty, the NPT, and the CWC demonstrate the degree to which there is such an underlying structure regardless of whether the behavior or activity is strictly military or has essentially civilian dimensions, whether all relevant states participate or only some of the most important states agree from the beginning to participate, and whether the agreement is global or regional in scope.

No verification system is entirely international; there is always some national element in the regime. Many multilateral agreements establish international procedures for addressing compliance issues among the parties, as do the Biological Weapons Convention and the Conventional Forces in Europe Treaty. Other treaties establish more elaborate international machinery to perform the routine (and some nonroutine) data collection and information analysis functions. This international machinery may even be delegated the authority to make judgments concerning compliance and noncompliance. However, governments always reserve for themselves some element of the compliance determination. The IAEA performs safeguards and is responsible for judging compliance with safeguards agreements, but compliance with the Nuclear Non-Proliferation Treaty is not judged by the IAEA. That authority belongs to the states parties, and to the United Nations Security Council (which in fact has the authority to judge compliance with any security related treaty and any behavior by one state with respect to potential threats to other states).[1] Likewise, in the Chemical Weapons Convention, the OPCW will perform the inspection and analysis functions, and the Executive Council (like the IAEA Board of Governors) takes action against parties not complying with their obligations under the CWC. In each case some element of judgment concerning the compliance of other parties (and the wisdom of responsive actions) is reserved to each state individually.

This finding is consistent with the model developed in chapter 1. States enter into multilateral agreements on arms control and disarmament (nonproliferation) for reasons of national security; the judgment is made that the national security is better protected by cooperating with other states in a treaty regime than it can be (or will be) by taking the unilateral alternative (that is, foregoing the treaty regime, which implies military means of deterrence and/or defense). There is something inherently contradictory in the idea that national security will be protected by ceding some sovereignty to an international organization. Sovereignty will be ceded only to the degree that benefits appear concrete and significant.

While this is clearly the case for major powers, small and weak states view this matter very differently. For them, ceding sovereignty is not a problem, for they may not be able to protect it in any case. They may actually prefer a multilateral treaty regime with intrusive verification, through which they can gain some measure of security for a cost they may have to pay in any case—that is, absent protection against weapons of mass destruction, they might ultimately lose more sovereignty to a hegomonic neighbor or regional power. This line of reasoning leads ultimately in the direction of supranational[2] organizations and political integration of states into larger units.[3] Security may be substantially increased by ceding sovereignty to a supranational organization, but at the cost of lost identity. States take such a step in the interest of the nations they govern only in the rarest of circumstances.

It is clear that the policy problem shapes the need for verification, but does not influence the shape of the verification system except in unusual cases (for example, if the real policy problem is comity, and the subject of agreement is less significant, there is little need for verification). In essence, a multilateral agreement presents different verification requirements than a bilateral agreement, whether the parties to a bilateral agreement are states or alliances. Factors driving the choice between establishing an international entity to perform verification and charging each state party with that responsibility would include the number of parties to the agreement and technical aspects, but not the particular type of policy problem. While the policy problem motivating the treaty will vary widely, the verification question is always the same: are other parties gaining some advantage by cheating on the mutual obligation?

Technical characteristics of the subject matter are normally a major factor in shaping the verification system. The nuclear industry was from the beginning heavily regulated (by national authorities), and accounting for nuclear materials is relatively straightforward. The international community had both the means of and a model for international "regulation" of national behavior when negotiating the NPT.[4] The BWC does not contain verification measures partly because of the difficulties in distinguishing between permitted and prohibited activities, and partly because of the judgment that it was unnecessary (as cheating would not lead to an effective battlefield weapon). But technical characteristics do not provide a means for explaining the striking differences between the CWC and the CFE Treaty (which raises the question of whether the IAEA/NPT verification approach is actually due to the factors just presented).

The factor that clearly plays a major role in the shape of each verification system is historical contingency. The first proposal for controlling the spread of nuclear weapons emphasized (more heavily than would any subsequent proposal for any of the cases addressed here) the creation of a single international organization which would be responsible for controlling the technology and verifying national compliance. Subsequent efforts to control nuclear weapons all began from that intellectual foundation; national verification was never considered for "horizontal" proliferation but only for "vertical" proliferation (an essentially bilateral, not multilateral, situation). When the BWC was negotiated, the British proposal included measures for international verification, but this approach was rejected in favor of one not including verification measures. The CWC grew out of the same early discussions as the BWC, but for chemical weapons a consensus existed that some verification system was essential. The question for negotiators was how would an international entity perform that verification, not whether verification would be the responsibility of individual states' parties or delegated to some international entity. Only in the CFE agreement was the national approach adopted, and in this instance, with the history of CSCE agreements in mind, the possibility of using one international entity to perform verification was never seriously considered. The fact that CFE was originally a "bilateral negotiation" strengthened this propensity.

In the end it appears that three factors are important at the macro level in determining whether verification will be performed by the parties individually or by an international entity specifically charged with that responsibility. The first is the number of parties expected to participate in the agreement: the more parties, the greater the costs of national verification and the more efficient international verification will be. Second is to some degree a question of the policy problem and to some degree a technical matter: is verification practical and relevant to the real versus the putative policy problem? Finally, the historical context in which a particular agreement is reached appears critical. For the NPT there was never any question—the international inspection system was already operating. The CWC, plus BWC discussions now going on, have always been heavily influenced by the NPT model, unlike the CFE treaty, which instead built on a model much closer in terms of political pedigree and the policy problem being addressed.

When one examines the finer grain of how verification is performed, the technical characteristics of the problem may become dominant, but here, too, historical contingency appears to play a relatively large role. The verification system eventually adopted usually makes sense in terms of the technical characteristics of prohibited and permitted activities, but one can identify alternative approaches that would have provided similar levels of confidence in the compliance judgments produced by verification.

There are lessons in this for negotiators of future treaties. The obvious lesson is that it is essential to know what the real policy problem is for each of the states participating in the negotiations. Different views on the type of policy problem will heavily influence judgments about the kind of appropriate verification measures.

The more subtle problem is to identify the historical contingencies that may shape the perceptions of negotiators from all states. Are the models "obviously" relevant in the perceptions of policy makers and negotiators in fact the models most appropriate for the problem at hand? In historical terms, the BWC might have been a better treaty if the negotiators had pursued a bilateral verification approach further rather than just requiring that parties assist in the investigation of "concerns" identified by other parties. Historical contingency defines the "problem view" of the negotiators and of the political and public audiences to be satisfied.

All four agreements studied were negotiated in a bipolar world of confrontational politics. The next major arms control or disarmament agreement, whether it be a comprehensive test ban treaty or a fissile materials production cutoff convention, will be negotiated in a very different political context. How definitions of the policy problem are shaped by this fundamental change, this largest of modern historical contingencies, cannot be gauged at this writing. Even the course of events in the NPT Extension Conference were predicated on very different political assumptions than was the original negotiation of the NPT.

It is difficult to step outside the historical moment in which one lives and see choices that do not fit in the existing frame of reference. To the degree negotiators are able to so, however, as did those who prepared the Acheson-Lillienthal report, they can shape history in a larger way than is otherwise possible.

Notes

1. See articles 24, 34, and 39 of the *Charter of the United Nations*, 26 June 1945 (TIAS 5857, 6529, and 7739).

2. And a supranational organization is really just an international organization which has gained some degree of independent sovereignty, an existence separate from the collective of states which make it up. The United States Federal Government is one example, the European Union is

another much more recent example.

3. While the issue was not weapons of mass destruction but simply overwhelming conventional power, it is this process which is taking place in Europe today. The European Union has its roots in the European Coal and Steel Community, which then became the European Economic Community. Both organizations had as their clear and well understood political objective the resolution of a security concern. Germany had become so overwhelmingly powerful in both economic and military terms that the only means available to her European neighbors to control that power was to intertwine their economies inextricably. As demonstrated by the recent Maastricht treaty, the ultimate product of this effort may well be a new United States of Europe.

4. It is worth noting that the English term "safeguards" is, in many other languages, the same term used for what we call "regulation." This use of terms has caused considerable confusion in international discourse, as "safeguards" and "regulation" are conceptually very different (although Americans may tend to overemphasize the differences just as many continental Europeans and others tend towards failing to fully appreciate the differences).

❖ Bibliography

Verification—Commentaries

Davis, Richard C., Lewis A. Dunn, Sidney Graybeal, Ralph Hallenbeck, Patricia McFate, and Timothy Pounds. *Arms Control Verification: Looking Back and Looking Ahead.* SAIC for the U.S. Department of Energy, June 22, 1993.

McFate, Patricia Bliss, "Where Do We Go From Here? Verifying Future Arms Control Agreements." *The Washington Quarterly* 15, no. 4 (Autumn 1992): 75-86.

Roberts, Brad, "Arms Control and the End of the Cold War." *The Washington Quarterly* 15, no.4 (Autumn 1992): 39-56.

Robinson, Julian Perry, *Chemical Warfare Arms Control: A framework for considering policy alternatives*, SIPRI Chemical & Biological Warfare Studies 2. London & Philadelphia: Taylor & Francis, 1985.

Sands, Philippe, "Enforcing Environmental Security: The Challenges of Compliance with International Obligations." *Journal of International Affairs* 46, no. 2 (Winter 1993).

Ziemki, Caroline, "Peace Without Strings? Interwar Naval Arms Control Revisited." *The Washington Quarterly* 15, no. 4 (Autumn 1992): 87-106.

Nuclear Weapons Proliferation—Texts

Statute of the International Atomic Energy Agency ("Statute"), as amended up to 28 December 1989 (International Atomic Energy Agency reprint).

Treaty for the Prohibition of Nuclear Weapons in Latin America ("Treaty of Tlatelolco"). Reprinted in U.S. Arms Control and Disarmament Agency, *Arms Control and Disarmament Agreements*, 68 - 86.

Treaty on the Non-Proliferation of Nuclear Weapons ("NPT"). Reprinted in United States Arms Control and Disarmament Agency, *Arms Control and Disarmament Agreements*, 98-102.

Information Circular 153, *The Structure and Content of Agreements Between the Agency and States Required in Connection with the Treaty on the Non-Proliferation of Nuclear Weapons*. International Atomic Energy Agency, June 1972.

Nuclear Weapons Proliferation—Commentaries

Blix, Hans. "Verification of Nuclear Nonproliferation: The Lessons of Iraq." *The Washington Quarterly* 15, no. 4 (Autumn 1992): 57-66.

Ekeus, Rolf. "The Iraqi Experience and the Future of Nuclear Nonproliferation." *The Washington Quarterly* 15, no. 4 (Autumn 1992): 67-74.

William Epstein. *The Last Chance: Nuclear Proliferation and Arms Control*. New York: The Free Press, 1976.

Fischer, David, Ben Sanders, Lawrence Scheinman, and George Bunn. *A New Nuclear Triad: The Non-Proliferation of Nuclear Weapons, International Verification and the International Atomic Energy Agency*. Programme for Promoting Nuclear Non-Proliferation PPNN Study Three, September 1992.

Fischer, David. "International Safeguards." In *Safeguarding the Atom: A Critical Appraisal*, ed. Jozef Goldblat. London & Philadelphia: Taylor & Francis for Stockholm International Peace Research Institute, 1985.

Kessler, J. Christian. "History & Current Trends in Nuclear Safeguards." Unpublished manuscript prepared for International

Training Course on Implementation of State Systems of Accounting for and Control of Nuclear Material, May 12 - 28, 1993. Sante Fe, New Mexico.

McKnight, Allan. *Atomic Safeguards: A Study in International Verification*. New York: United Nations Institute for Training and Research, 1971.

Quester, George H., and Victor A. Utgoff. "U.S. Arms Reductions and Nuclear Nonproliferation: The Counterproductive Possibilities." *The Washington Quarterly* 16, no. 1 (Winter 1993): 129-140.

Scheinman, Lawrence. *The International Atomic Energy Agency and World Nuclear Order*. Washington, DC: Resources for the Future, 1987.

____________. "Safeguards: New Threats and New Expectations." *Disarmament: A periodic review by the United Nations* XV, no. 2, 1992.

____________. "Lessons From Post-War Iraq for the International Full-Scope Safeguards Regime." *Arms Control Today* 23, no. 3: 3-6.

Trevan, Tim. "UNSCOM Faces Entirely New Verification Challenges in Iraq." *Arms Control Today* 23, no. 3: 11-15.

Van Doren, Charles N., and George Bunn, "Progress and Peril at the Fourth NPT Review Conference. *Arms Control Today* 20, no. 8 (October 1990): 8-12.

Zifferero, Maurizio. "The IAEA: Neutralizing Iraq's Nuclear Weapons Potential." *Arms Control Today* 23, no. 3: 7-10.

Biological Weapons Proliferation—Texts

Protocol for the Prohibition of the Use in War of Asphyxiating, Poisonous or Other Gases, and of Bacteriological Methods of

Warfare (" Geneva Protocol"). Reprinted in United States Arms Control and Disarmament Agency, *Arms Control and Disarmament Agreements*, 15-19.

Convention on the Prohibition of the Development, Production and Stockpiling of Bacteriological (Biological) and Toxin Weapons and on Their Destruction ("Biological Weapons Convention"). Reprinted in United States Arms Control and Disarmament Agency, *Arms Control and Disarmament Agreements*, 133-137.

"Final Declaration of the First Review Conference of the Parties to the Biological Weapons Convention, March 1980." Reprinted in *Preventing a Biological Arms Race*, ed. Susan Wright. Cambridge, MA: The MIT Press, 1990.

"Final Declaration of the Second Review Conference of the Parties to the Biological Weapons Convention, September 1986." Reprinted in *Preventing a Biological Arms Race*, ed. Susan Wright. Cambridge, MA: The MIT Press, 1990.

"Final Declaration." Part II of *Third Review Conference of the Parties to the Convention on the Prohibition of the Development, Production and Stockpiling of Bacteriological (Biological) and Toxin Weapons and on their Destruction (Geneva, 9 - 27 September 1991). FINAL DOCUMENT*, BWC/CONF.III/23.

Summary of the work of the Ad Hoc Group for the period 23 November to 4 December 1992. Ad hoc Group of Governmental Experts to Identify and Examine Potential Verification Measures from a Scientific and Technical Standpoint. BWC/CONF.III/VEREX/4, 8 December 1992

United Kingdom, Verification of the BWC: Possible Directions. Ad hoc Group of Governmental Experts to Identify and Examine Potential Verification Measures from a Scientific and Technical Standpoint. BWC/CONF.III/VEREX/WP.1, 30 March 1992 *Summary Report*. Ad hoc Group of Governmental Experts to Identify and Examine Potential Verification Measures from a Scientific and Technical

Standpoint ("VEREX Report"). BWC/CONF.III/VEREX/8, Geneva, 24 September 1993.

Biological Weapons Proliferation—Commentaries

Cole, Leonard. "Sverdlovsk, Yellow Rain, and Novel Soviet Bioweapons: Allegations and Responses." In *Preventing a Biological Arms Race*, ed. Susan Wright. Cambridge, MA: The MIT Press, 1990.

Falk, Richard. "Inhibiting Reliance on Biological Weaponry: The Role and Relevance of International Law." In *Preventing a Biological Arms Race*, Cambridge, MA: The MIT Press, 1990.

Falk, Richard, and Susan Wright. "Preventing a Biological Arms Race: New Initiatives," in Susan Wright (ed.), *Preventing a Biological Arms Race*, ed. Susan Wright. Cambridge, MA: The MIT Press, 1990.

Leitenberg, Milton. "Anthrax in Sverdlovsk: New Pieces to the Puzzle." *Arms Control Today* 22, no. 3 (April 1992): 10-13.

Pearson, Graham S. "Prospects for Chemical and Biological Arms Control: The Web of Deterrence." *The Washington Quarterly* 16, no. 2 (Spring 1993): 145-162.

Robinson, Julian, Jeanne Guillemin, and Matthew Meselson. "Yellow Rain in Southeast Asia: The Story Collapses." In *Preventing a Biological Arms Race*, ed. Susan Wright. Cambridge, MA: The MIT Press, 1990.

Rosenberg, Barbara Hatch, and Gordon Burck. "Verification of Compliance with the Biological Weapons Convention." In *Preventing a Biological Arms Race*, ed. Susan Wright. Cambridge, MA: The MIT Press, 1990.

Sims, Nicholas. "The Second Review Conference on the Biological Weapons Convention." In *Preventing a Biological Arms Race*, ed. Susan Wright. Cambridge, MA: The MIT Press, 1990.

————————. *The Diplomacy of Biological Disarmament: Vicissitudes of a Treaty in Force, 1975-85* New York: St. Martin's Press, 1988

Wright, Susan. "Evolution of Biological Warfare Policy, 1945 - 1990." In *Preventing a Biological Arms Race*, ed. Susan Wright. Cambridge, MA: The MIT Press, 1990.

Chemical Weapons Proliferation— Texts

Protocol for the Prohibition of the Use in War of Asphyxiating, Poisonous or Other Gases, and of Bacteriological Methods of Warfare (" Geneva Protocol"). Reprinted in United States Arms Control and Disarmament Agency, *Arms Control and Disarmament Agreements*, 15-19.

Convention on the Prohibition of the Development, Production, Stockpiling and Use of Chemical Weapons and on their Destruction, Paris, January 13, 1993,.U.S. Arms Control and Disarmament Agency reprint, October 1993.

Chemical Weapons Proliferation—Commentaries

Aroesty, J., K. A. Wolf, and E. C. River. *Domestic Implementation of a Chemical Weapons Treaty*. Santa Monica, CA: RAND National Defense Research Institute, October 1989

Brin, Jay. "Ending the Scourge of Chemical Weapons." *Technology Review*, April 93, 30-40.

Carnahan, Burrus. "Constitutional Implications of Implementing a Chemical Weapons Convention." SAIC, April 4, 1990.

Carus, W. Seth. "The Proliferation of Chemical Weapons Without a Convention." In *Chemical Disarmament and U.S. Security*, ed. Brad Roberts. Boulder, CO: Westview Press, 1992.

Fahmy, Nabil. "The Security of Developing Countries and Chemical Disarmament." In *Chemical Disarmament and U.S. Security*, ed. Brad Roberts. Boulder, CO: Westview Press, 1992.

Feinstein, Lee. "Geneva Negotiators Complete Chemical Weapons Treaty." *Arms Control Today* 22, no. 7 (September 1992): 24, 30.

Flowerree, Charles C. "Implementing the Chemical Weapons Convention." In *Chemical Disarmament and U.S. Security*, ed. Brad Roberts. Boulder, CO: Westview Press, 1992.

Kirk, Roger. "Lessons of the IAEA Experience for a Chemical Weapons Verification Organization." SAIC, 26 December 1990.

Krause, Joachim, "The Changing International System and Chemical Disarmament." In *Chemical Disarmament and U.S. Security*, ed. Brad Roberts. Boulder, CO: Westview Press, 1992.

Krepon, Michael. "Verification of a Chemical Weapons Convention." In *Chemical Disarmament and U.S. Security*, ed. Brad Roberts. Boulder, CO: Westview Press, 1992.

Ledogar, Ambassador Stephen J. Ledogar. "Closing in on a Chemical Weapons Ban." Interview in *Arms Control Today* 21, no. 4 (May 1991): 3-7.

Lehman, Ronald F. "Concluding the Chemical Weapons Convention." In *Chemical Disarmament and U.S. Security*, ed. Brad Roberts. Boulder, CO: Westview Press, 1992.

Lancaster, H. Martin. "The Future of U.S. Chemical Warfare Capabilities."In *Chemical Disarmament and U.S. Security*, ed. Brad Roberts. Boulder, CO: Westview Press, 1992.

Mullen, Mark. "Verification of a Chemical Weapons Convention: Summary of Lessons Learned from the Verification Experience of the International Atomic Energy Agency." Los Alamos National Laboratory LA-UR-90-2674.

Olson, Kyle. "Disarmament and the Chemical Industry." In *Chemical Disarmament and U.S. Security*, ed. Brad Roberts. Boulder, CO: Westview Press, 1992.

Pearson, Graham, S. "Prospects for Chemical and Biological Arms Control: The Web of Deterrence" (see ***Biological Weapons Proliferation—Commentaries***).

Report: Workshop on the Experiences of the International Atomic Energy Agency (IAEA) as a Model for the Organization for the Prohibition of Chemical Weapons. Meridian Corporation (January 28, 1993), undated.

Robinson, Julian Perry. *Chemical Warfare Arms Control: A framework for considering policy alternatives.* SIPRI Chemical & Biological Warfare Studies 2. London & Philadelphia: Taylor & Francis, 1985

Roberts, Brad. "Framing the Ratification Debate." In *Chemical Disarmament and U.S. Security*, ed. Brad Roberts. Boulder, CO: Westview Press, 1992.

Smithson, Amy E. ed., *Administering the Chemical Weapons Convention: Lessons from the IAEA.* The Henry L. Stimson Center Occasional Paper No. 14, March 1993.

Utgoff, Victor, and Susan Leibbrandt. "On the Pursuit of Universal Adherence to the Chemical Weapons Convention." In *Chemical Disarmament and U.S. Security*, ed. Brad Roberts. Boulder, CO: Westview Press, 1992.

U.S. Arms Control and Disarmament Agency. *Fact Sheet, Chemical Weapons Negotiations at the Conference on Disarmament*, August 13, 1992.

Walker, John, "Challenge Inspections and Intrusiveness." In *Chemical Disarmament and U.S. Security*, ed. Brad Roberts. Boulder, CO: Westview Press, 1992.

Wilson, Trevor. "Chemical Disarmament versus Chemical Nonproliferation." In *Chemical Disarmament and U.S. Security*, ed. Brad Roberts. Boulder, CO: Westview Press, 1992.

Conventional Forces in Europe—Texts

Document of the Stockholm Conference on Confidence- and Security-Building Measures and Disarmament in Europe Convened in Accordance With the Relevant Provisions of the Concluding Document of the Madrid Meeting of the Conference on Security and Cooperation in Europe. Reprinted in United States Arms Control and Disarmament Agency, *Arms Control and Disarmament Agreements*, 323 - 335.

Treaty on Conventional Armed Forced in Europe, Paris, November 19, 1990 (U.S. Arms Control and Disarmament Agency, printed by Regional Program Office Vienna, United States Information Agency).

Conventional Forces in Europe—Commentaries

Feinstein, Lee. "Superpowers Recast Proposals as CFE Deadline Nears." *Arms Control Today* 20, no. 8 (October 1990): 25, 28.

____________. "CFE Terms Tentatively Agreed; Signing Expected in November." *Arms Control Today* 20, no. 8 (October 1990): 24, 27.

____________. "Bush and Gorbachev Move Toward Compromise on CFE." *Arms Control Today* 21, no. 4 (May 1991): 20.

____________. "25 Nations Sign CFE Follow-on." *Arms Control Today* 22, no. 6 (July/August 1992): 29.

____________. "Russia Asks CFE Partners to Allow Increase in Arms on Southern Border." *Arms Control Today* 23, no. 9 (November 1993):25.

______________. "CFE Executive Summary." *Arms Control Today* 21, no. 1 (January/February 1991): "CFE Supplement."

______________. "CFE: Off the Endangered List?." *Arms Control Today* 23, no. 8 (October 1993): 3-6.

Graham, Thomas, Jr.. "The CFE Story: Tales From the Negotiating Table." *Arms Control Today*, European Security Special Issue, 22, no. 1 (January/February 1991): 9-11.

U.S. Arms Control and Disarmament Agency *Fact Sheet.* "A Chronology of Arms Control and Related Treaties and Agreements including Confidence- and Security-Building Measures, and Measures Related to Non-Proliferation, Transparency and Defense Conversion." December 20, 1993.

Comprehensive Test Ban—Commentaries

"Ambassador C. Paul Robinson: Verifying Testing Treaties—Old and New." Interview in *Arms Control Today* 20, no. 6 (July/August 1990): 3-7.

Kaplow, David A. "The Step-by-Step Approach." *Arms Control Today* CTB Special Issue, 20, no. 9 (November 1990): 3, 6-8.

Kidder, Ray E. "How Much More Nuclear Testing Do We Need?" *Arms Control Today* 22, no. 7 (September, 1992): 11-14.

Mark, J. Carson. "Do We Need Nuclear Testing?" *Arms Control Today* 22, no. 7 (September, 1992): 12-17.

Mendelsohn, Jack. "Arms Control After the Summit." *Arms Control Today* 23, no. 4 (May 1993): 10-13.

Panofsky, Wolfgang K. H. "Verification of the Threshold Test Ban." *Arms Control Today* 20, no. 7 (September 1990):3-6.

____________. "Straight to a CTB." *Arms Control Today* CTB Special Issue, 20, no. 9 (November 1990): 3-5.

Smith, Gerard C. "End Testing, Stem the Bomb's Spread." *Arms Control Today* CTB Special Issue, 20, no. 9 (November 1990): 9-11.

van der Vink, Gregory E. "Verifying a Comprehensive Test Ban." *Arms Control Today* CTB Special Issue, 20, no. 9 (November 1990): 18-23.

Fissile Materials Production Cutoff—Commentaries

Mendelsohn, Jack, "Arms Control After the Summit" (see ***Comprehensive Test Ban Treaty—Commentaries***).

❖ About the Author

J. Christian Kessler is deputy director of the Office of Regional Non-Proliferation, Political-Military Affairs Bureau, Department of State, where he is responsible for U.S. regional security policies to prevent the proliferation of nuclear, chemical, and biological weapons and missile delivery systems. In this capacity, Mr. Kessler helped supervise Project Sapphire, in which the United States removed 600 kilograms of weapons-usable nuclear material from Kazakhstan. Previously he was director of the Safeguards and Safety Division, State Department Office of Nuclear Technology & Safeguards. Prior to joining the State Department, Mr. Kessler worked for the U.S. Nuclear Regulatory Commission on international safeguards and export licensing matters.

Mr. Kessler received his undergraduate degree from Bowdoin College and holds master' degrees from Indiana University and the National War College. He is the author of numerous articles on international nuclear safeguards and has published on terrorism and maritime industrial security issues.

*U.S. G.P.O.:1995-387-330:20010